MECHANICAL FASTENING
OF PLASTICS

MECHANICAL ENGINEERING

A Series of Textbooks and Reference Books

EDITORS

L. L. FAULKNER

Department of Mechanical Engineering
The Ohio State University
Columbus, Ohio

S. B. MENKES

Department of Mechanical Engineering
The City College of the
City University of New York
New York, New York

1. Spring Designer's Handbook, *by Harold Carlson*
2. Computer-Aided Graphics and Design, *by Daniel L. Ryan*
3. Lubrication Fundamentals, *by J. George Wills*
4. Solar Engineering for Domestic Buildings, *by William A. Himmelman*
5. Applied Engineering Mechanics: Statics and Dynamics, *by G. Boothroyd and C. Poli*
6. Centrifugal Pump Clinic, *by Igor J. Karassik*
7. Computer-Aided Kinetics for Machine Design, *by Daniel L. Ryan*
8. Plastics Products Design Handbook, Part A: Materials and Components; Part B: Processes and Design for Processes, *edited by Edward Miller*
9. Turbomachinery: Basic Theory and Applications, *by Earl Logan, Jr.*
10. Vibrations of Shells and Plates, *by Werner Soedel*
11. Flat and Corrugated Diaphragm Design Handbook, *by Mario Di Giovanni*
12. Practical Stress Analysis in Engineering Design, *by Alexander Blake*
13. An Introduction to the Design and Behavior of Bolted Joints, *by John H. Bickford*
14. Optimal Engineering Design: Principles and Applications, *by James N. Siddall*
15. Spring Manufacturing Handbook, *by Harold Carlson*

OTHER VOLUMES IN PREPARATION

MECHANICAL FASTENING OF PLASTICS
OF PLASTICS
An Engineering Handbook

Brayton Lincoln
Kenneth J. Gomes
James F. Braden

MARCEL DEKKER, INC.   New York and Basel

Library of Congress Cataloging in Publication Data

Lincoln, Brayton, [date]
  Mechanical fastening of plastics.

  (Mechanical engineering; 26)
  Includes index.
  1. Fasteners--Handbooks, manuals, etc. 2. Plastics--
Handbooks, manuals, etc. I. Gomes, Kenneth J.,
[date]. II. Braden, James F., [date].
III. Title. IV. Series.
TJ1320.L48   1984   621.8'8'028   83-19013
0-8247-7078-1

MARCEL DEKKER, INC.
270 Madison Avenue, New York, New York 10016

Current printing (last digit):
10 9 8 7 6 5 4 3

PRINTED IN THE UNITED STATES OF AMERICA

To
George A. Andrews
Arthur D. Bancroft
Arthur E. Schneider
Robert L. Welter

PREFACE

Robotics, automation, electronics, communication, plastics—these are some of the buzzwords of the world's current industrial directions. At the 1982 National Design Engineering Show approximately 30 percent of respondents to a questionnaire advised that they expected their current or new designs to encompass robotic assembly in one way or another. It is estimated that by the mid-1990s automation, made possible through electronic developments, will have displaced from industry all but 10 percent of our productive workforce, down from approximately 22 percent in the early 1980s.

For those of us engaged in manufacturing and/or marketing of such fundamental products as fasteners and tools, or in designing and assembling new components, the substitution of an electronic and communications base for traditional leading markets such as steel, automotive, building construction, and related industries is having profound consequences.

At the same time that the new industrial direction has been developing, it has been accompanied by a revolution in the manufacturing materials being used—and at about the same pace. This accompanying revolution is of course the switch from the use of metals to the use of plastics. It is very evident in many fields. The use of plastics has been expanding rapidly; over the past 20 years, there has been a tremendous expansion in the development of new ones—harder, more structural polymers and blends. Many plastics are hitting hard at specific uses of metals. More than 40 "families" of plastics are in current use. Each can have dozens of variable subgroups, all with different fastening characteristics.

The designer of plastic parts, already faced with a bewildering series of possible polymer and filler combinations, may face even more difficulties in the future because of possible material property overlaps.

In the field of metals there is a wealth of knowledge covering mechanical fastening, and yet, in spite of all the research that has been done over the past 100 years, not everything about fastening in metals has been researched or covered. In the field of plastics, because of the tremendous surge in material development, the research surface has hardly been scratched. Not much has been published about fastening plastics.

It seems quite evident, therefore, that for those of us now engaged or about to be engaged with the assembly and fastening of products, such as product designers, manufacturing engineers, buyers, fastener marketing personnel, and students, there is a pressing need to understand the new materials and how fasteners work in them and an accompanying requirement to understand how fastening methods will be affected by switches to automatic assembly. This book endeavors to be a modest beginning to the creation of such an understanding.

Brayton Lincoln
Kenneth J. Gomes
James F. Braden

ACKNOWLEDGMENTS

We are especially indebted to Continental/Midland, a unit of AMCA International Corporation for the generous permission to make copious use of material from their considerable publications, laboratory studies, and audiovisual programs previously developed by the authors while they were employed at Continental Screw on the subject of fastening in plastics. Most of the illustrations in Chapters 2, 7, and 8 are in this category.

The general ideas for Sec. 1.1, Chap. 1, "Organization of the Plastics Goods Manufacturing Process, came from *Plastic Design Forum's* published article "The Plastics Production Process," © 1980 Industry Media, Inc., Denver, Colorado. The authors wish to express their thanks to the publisher for permission to incorporate these ideas in this volume.

Many of the core ideas of this book were previously published in two articles by Kenneth J. Gomes in the August and September 1982 issues of *Assembly Engineering*, under the title "Fastening Plastics," © 1982, Hitchcock Publishing Co., Inc., Wheaton, Illinois. The authors gratefully acknowledge permission of the publisher to use illustrations and quote extensively from these articles.

Special thanks to:

AKKO Fastener
  Bob Adler
  Ken Delape

L. & C. Arnold
  Werner C. Berger
  Manfred Schwarz

ARO Corp.
  Phillips Milan
  Thomas L. Reed

Assembly Engineering Magazine
  Mary Emrich
  Robert J. Kelly
  Roland Laboissonierre
  Terry Thompson

Black and Webster
  Bob Dennen
  Peter G. MacLaren
Bulten-Newall
  Malcolm MacGregor
Camcar/Textron
  Bernard F. Reiland
Carr Division of TRW, Inc.
  Donald R. Portlock
Conti, A. G.
  Barry Hughes
Continental/Midland
  Claire Acheson
  George A. Andrews
  Arthur D. Bancroft
  Dennis Boyer
  Rick Caravalho
  Eugene Chapman
  H. Kurt Dieme
  Pat Graham
  Herman Muenchinger
  Louis G. Roehr
  Steven Rooney
  Doris Sanford
  Arthur E. Schneider
  David Walker
  Robert L. Welter
DuPont
  James H. Crate
ELCO Industries
  Joe Kordash
  Jack Rolf
General Electric Plastics,
  Pittsfield
  Renee Miller
  Sam Miller
  Edward Stumpek
GKN Screws & Fasteners, Ltd.
  W. John Heap

Gobin Daude, S. A.
  Max Augenthaler
Greenfield Tap & Die Div., TRW
  Dino Emmanuele
Helicoil, Div. Mite Corp.
  Gary Canestrari
  Bill Downing
  Al Mussnug
  Jim Wilson
Japan Fastener Research and
  Engineering Co., Ltd.
  Akira Kasai
  Marty Kishiro
Linread, Ltd.
  Donald G. Lynall
  John Parker
Modern Plastics Encyclopedia
  Joan Agranoff
Plastics Design Forum
  Deb Brewer
The Polymer Corporation
  Leroy Q. Wenrich
PSM Fasteners, Ltd.
  Jimmy Tildesley
Reed and Prince Mfg. Co.
  Gil Rainsford
Sonics and Materials, Inc.
  Robert Armstrong
  Michael Donnaty
U.S. Army R & D Command, Plastics
  Technical Evaluation Center,
  Picatinny Arsenal
  John Nardonne
Weber Automatic Screw Drivers and
  Assembly Systems, Inc.
  John Grey
  Dieter Uckert

CONTENTS

Chapter 5 Tradename Plastics   91

MECHANICAL FASTENING
OF PLASTICS

1
INTRODUCTION TO PLASTICS

The use of plastics has become widespread throughout industry, replacing aluminum and zinc die castings and other metal parts in virtually every category of product design. In 1981, it was estimated that 20% of the diecast market had transferred to plastics (1). This trend is expected to continue for several reasons, in particular the cost differential between molding plastics and forming metals, even though the newer replacements may be closer to the functional limitations of plastics.

Although the material cost of the raw plastic material may be higher than that of metals, the processing cost is generally lower, so that, overall, parts can be made more economically. In some cases, perhaps because of volume considerations, the cost of a plastic part may exceed the cost of metal. Nevertheless, other considerations, such as weight differential or corrosion resistance, may be deciding factors in the selection of plastics over metals.

A favorite example of lecturers and writers on the expanded use of plastics has been the automobile. In 1971, it was estimated that around 100 pounds of plastics were being used per car. A couple of years ago it was projected that by 1990 300 pounds of plastics per car would be used. There is now more reticence in citing estimates on autos because of the rapid increase in the actual and experimental usages of plastics in automobile manufacturing. It is not entirely facetious to remark that soon there will be nothing left to convert but the passengers, and we are not sure how much plastic they will contain.

Also pertinent to the increased usage of plastics is the continuing development of new and stronger combinations of materials and fillers and the development of a broader understanding of how to work with plastics. The increased usage of structurally strong plastics has presented more opportunities to use mechanical fasteners that meet the demands of the new materials.

1

In the past, an understanding of fastening in plastics was difficult to obtain. Part of the difficulty was due to the proliferation of grades, types, combinations, and possible quantity variations of fillers. In our talks on the subject of fastening in plastics we have been using the figure of 4500 variations. Perhaps this figure is somewhat exaggerated. In any case, because of the large number of materials and filler combinations, each with different fastening characteristics, care must be taken to match the correct fastener to the individual application.

Of course, design engineers rarely enjoy the luxury of selecting a fastening method at the planning or drawing board stage. Functional and environmental requirements take top priority, and chemical, mechanical, and electrical characteristics, as well as manufacturing, material costs, and cosmetic considerations will undoubtedly have to be considered in material selection—thus, design of the fastening site and selection of the fastener are usually last to be considered. Furthermore, in the past there has been little information on fastening in plastics, and what was available was sometimes misleading.

1.1 ORGANIZATION OF THE PLASTIC GOODS MANUFACTURING PROCESS

Because of the wide variety of plastic materials on the market today and the range of operations involved, the manufacturing process for plastic goods has become fairly complex (2). It may seem to the outsider that there is no clear chain of command. For those of us who are newly interested in the subject, and at the risk of oversimplification we will attempt to show how the manufacturing process works (Fig. 1.1).

1.1.1 Resin Producers

Resin producers are the chemical organizations that supply the basic plastics materials or polymers from which the finished goods will be manufactured. They are generally rather large companies, including Du Pont, Union Carbide, General Electric, and Monsanto with large technical staffs that are usually enthusiastic about promoting their resins by providing technical assistance and advice. The polymers that they produce can be in liquid, pellet, granule, powder, or other form. This material is supplied to the compounder.

1.1.2 Compounders

Compounders may be part of the resin producer's organization (captive compounders) or they may be independent compounders. In either case, the compounder may further modify or enhance the basic resin by adding or blending other materials with it. Such modifications could

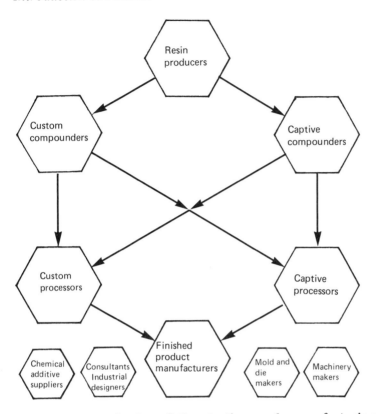

FIG. 1.1 Organization of the plastics goods manufacturing process.
(Adapted from "The Plastics Production Process," *Plastics Design Forum* ©1980 Industry Media, Inc., Denver, CO)

be as simple as mixing in filler such as minerals or glass, in powdered or fiber form.

1.1.3 Processors

The processor's job is to form the compounded material into a finished shape or part by molding, extruding, laminating, or other method. The processor can also be a captive organization or an independent contractor. They may or may not be doing the work for their own use. They work closely with the designer, the finished goods manufacturer, and the compounder.

1.1.4 Finished Goods Manufacturers

The manufacturer of the finished product produces the completed product for sale. As part of the operation, the manufacturer may also decorate and/or machine the product and may also do the processing. The designer of the product may or may not be a direct employee. In either case, the designer, or design consultant, is at the heart of the process. He or she will probably be working with processing engineers who can help in the material manufacturing process and the material selection, production methods, finishing, and so forth.

Correlation of finished product design with the assembly and processing techniques to be employed is an obvious requirement in order to produce the product at the optimum cost level. Less obvious to the newcomer, but of equal importance, is the correlation of material with the processing technique to be used in the molding. However, it has been conclusively demonstrated that the finished material's physical and mechanical characteristics are heavily influenced by the molding techniques employed.

1.1.5 Suppliers

Also involved in the manufacturing process are mold and die makers, process machinery manufacturers, chemical additive suppliers, and many other suppliers (2), including the fastener manufacturer, who will be working with the designer and later with the assembly engineers.

1.2 PLASTICS VERSUS METALS

The decision to use plastic in place of metal is not always clear-cut and should be made only after a detailed comparison of the many accounting and physical requirements involved (3). For instance, if disassembly is involved, then consideration should be given to fastening-site design at this point. This factor would differ with the use of metals.

It is not the purpose of this volume to discuss material selection in detail, but for the benefit of some readers it should be noted that if the necessary physical requirements can be met with plastic materials, then cost factors regarding secondary operations required for metal parts may well favor selection of plastics (3) (Fig. 1.2). It will be noted from this illustration that manufacturing operations are far fewer with plastics then with metals.

When making the choice between metals and plastics, the designer will have to consider required physical conditions including environment (temperature, chemical atmosphere, humidity, etc.), longevity, electrical conductivity or resistivity, weight, appearance, and tolerances. With plastics, the designer will have to consider required strength, creep resistance, dimensional stability, stress and strain curves, and loadings (4), perhaps even coefficient of friction. Considerably more care is required in selecting plastics for design use than metals as the structural properties of plastics are more sensitive to environmental factors. On the other hand, raw material considerations

---

Steps in diecasting

1. Transport ingot to furnace	13. Inspect for voids, porosity
2. Drop ingot into melt furnace	14. Transport part to next operation
3. Melt ingot; skim dross	15. Belt sand part
4. Transfer melted ingot into holding furnace	16. Drill and ream
5. Clean and inspect die	17. Tap
6. Flux and cool die	18. Face
7. Remove melted metal from holding furnace	19. Inspect for tolerances and so on
8. Ladle molten metal into shot chamber	20. Transport part to finishing area
9. Wait for part to cool	21. Tumble, deburr
10. Release part from die	22. Inspect surfaces
11. Set up for deflashing	23. Rack, degrease
12. Deflash	24. Paint, dry, unrack
	25. Box and ship

Steps in injection molding

1. Load resin into hopper
2. Cycle
3. Remove finished part
4. Box and ship

FIG. 1.2 Why die cast aluminum costs so much. (Courtesy of General Electric Company, Specialty Plastics Division, Pittsfield, MA. Reprinted from Plastics Design Forum, January/February, 1981, p. 35; copyright © Industry Media, Inc., Denver, CO, by permission of the publisher.)

involving factors of future availability, energy requirements, and inflationary trends probably favor selection of plastic.

The long-term trends regarding strength-to-weight comparisons also appear to favor selection of plastics in many instances, even though die castings have better dimensional stability, have closer tolerances, and are stiffer and stronger. It is becoming increasingly clear, as plastics engineers work with glass and other fiber fillers (including carbon), as well as mineral reinforcements, that composite-filled materials can be engineered or "tailored" to meet high-performance requirements (1), and that this can be done on a selected location basis.

In addition, energy requirements for plastic processing are also generally lower than those necessary for processing metals. For example, aluminum, which requires electrical energy for processing, may need three times as much energy for initial manufacturing as the comparative strength equivalent in plastic. Of course, energy requirements for reprocessing aluminum are considerably lower than for initial manufacture.

1.3 A BRIEF EXPLANATION OF PLASTICS
FOR THE FASTENER ENGINEER

Plastics are synthetic materials composed of a series or chain of molecules which, when heat or pressure is applied, can be formed into desired shapes. The eight elements used to manufacture plastics are hydrogen, carbon, nitrogen, oxygen, fluorine, silicon, sulfur, and chlorine. These elements can be bonded together in stable compounds in accordance with their individual energy-bond capacities.

The basic unit from which plastics are derived is called a *monomer*. Ethylene is a monomer which has the formula

$$C_2H_4 \quad \text{or} \quad \begin{matrix} H & H \\ | & | \\ C & = & C \\ | & | \\ H & H \end{matrix}$$

The monomer form is subjected to heat-pressure-catalytic processes to form chains of monomers which are called *polymers*. The length of the molecular chain partially determines the material characteristics. An ultrahigh-molecular-weight polyethylene (UHMWPE) chain can be thousands of units long. An ideal polyethylene chain structure is designated:

$$(CH_2 - CH_2)_n \quad \text{or} \quad \begin{matrix} H & H \\ | & | \\ -(C - C)_n - \\ | & | \\ H & H \end{matrix}$$

A polymer made from a single monomer is called a *homopolymer*. A polymer made from two different monomers is called a *copolymer*. A copolymer has different properties than a homopolymer made from either of the two monomers that compose it. The combination of properties will depend on the processing conditions (5), and the resulting properties will be considerably affected by the structure (6). An example of a copolymer is polyvinyl chloride (PVC):

$$\text{(CH}_2 - \text{CHCl)}_n \quad \text{or} \quad -(\overset{\displaystyle H}{\underset{\displaystyle H}{\overset{|}{\underset{|}{C}}}} - \overset{\displaystyle H}{\underset{\displaystyle Cl}{\overset{|}{\underset{|}{C}}}})_n-$$

Terpolymers consist of three monomers. Acrylonitrile, butadiene, and styrene (ABS) resins form a terpolymer.

There are also blends or combinations of polymers that are mixed together mechanically without chemical bonding. These are sometimes called *alloys*.

In the manufacturing process, which takes place in a closed container or reactor, the required monomers are subjected to heat and pressure in the presence of catalytic substances. The catalysts assist in the molecular recombination of monomeric bondings into many polymeric chains. These grow in length simultaneously until all the uncombined monomers have combined with a free hydrogen (or other chain-stopping) atom at the both ends of the chain:

$$(H)-\overset{\displaystyle H}{\underset{\displaystyle H}{\overset{|}{\underset{|}{C}}}}-\overset{\displaystyle H}{\underset{\displaystyle H}{\overset{|}{\underset{|}{C}}}}-\overset{\displaystyle H}{\underset{\displaystyle H}{\overset{|}{\underset{|}{C}}}}-\overset{\displaystyle H}{\underset{\displaystyle H}{\overset{|}{\underset{|}{C}}}}-\overset{\displaystyle H}{\underset{\displaystyle H}{\overset{|}{\underset{|}{C}}}}-\overset{\displaystyle H}{\underset{\displaystyle H}{\overset{|}{\underset{|}{C}}}}-\overset{\displaystyle H}{\underset{\displaystyle H}{\overset{|}{\underset{|}{C}}}}-\overset{\displaystyle H}{\underset{\displaystyle H}{\overset{|}{\underset{|}{C}}}}-\overset{\displaystyle H}{\underset{\displaystyle H}{\overset{|}{\underset{|}{C}}}}-(H)$$

This chain is an ideal structure; during most manufacturing processing, different degrees and kinds of side branching will occur (7). The length of the polymer chain can be approximately controlled by predetermined addition of chain-stopping elements.

The longer the chain, the greater the molecular weight. Higher molecular weight increases toughness, resistance to creep, and stress cracking, but it also increases processing difficulty and cost (5).

These molecular chains are not straight but interwine with each other in a haphazard manner, often referred to as "spaghetti-like," sometimes as "wormlike," with small segments that are in constant motion. In polymer chemistry such spaghetti-like structure is called "amorphous." Some plastics have a "crystalline" or more-ordered internal structure; most of these are partially crystalline, partially amorphous; some plastics are wholly amorphous.

Although plastics chemistry is much more complicated than it appears from our explanation, it is the ability to combine monomer chains that allows the plastics industry to provide controlled and predictable properties in the end product. While each plastic offers special, unique qualities, these materials achieve even greater variety by the numerous forms in which they can be produced.

Whatever their properties or form, plastics generally fall into two groups: thermoplastics and thermosets, each with a number of subgroups.

1.4   THERMOPLASTICS VERSUS THERMOSET
(CROSS-LINKED) PLASTICS

At the risk of oversimplification, we will say that thermoplastics act
somewhat like ice and water. More correctly, they behave like highly
viscous liquids and/or viscoelastic solids. At usable temperatures
thermoplastics are solid, but at elevated temperatures they become
soft and melt. As the temperature is lowered, they will resolidify.
Crystalline thermoplastics have a rather sharp transition temperature.
With amorphous thermoplastics, the change will be over a range of temperatures. In contrast, thermoset plastics have a molecular, cross-
linked structure that is irrevocably "set" into permanent shape during
the manufacturing process.

The techniques used for processing thermoplastics are quite different from those used for thermoset plastics. Thermoplastics are completely prepared for use by the primary material producer and compounder, and the molder merely heats them to use. Thermosets, on
the other hand, are only partially prepared by the primary material
producer, who stops or restricts the action after formation of the polymer chains. Depending on the thermoset, completion of the process
(cross-linkage) is accomplished in the molding operation, either through
a condensation molding method or through the addition of required components that induce cross-linkage.

The intermolecular forces that hold thermoplastic polymeric chains
together weaken with heating and strengthen as they cool. However,
if too much heat is applied, the properties of thermoplastics can be
permanently degraded.

We have mentioned intermolecular forces which hold the chains together. In addition to the mechanically intertwined nature of the amorphous structure, there are bonds between the molecules within the chain
which hold each chain together as a unit. There is also a group of
secondary and not very powerful individual forces which supplement
the mechanical intertwining, but acting as a whole are sufficiently strong
to hold the polymer chains together as one amorphous unit.

In addition to chain length, polymer molecular shape and chain side-
branching are important to the properties of the plastic. The closer
the polymers can pack together, the higher the density of the plastic
(Fig. 1.3) and the higher the percentage of crystallinity. Higher
crystallinity tends to provide increased stiffness and resistance to
creep. At the same time it promotes less resistance to stress cracks.
Amorphous (noncrystalline) structure tends to be more flexible and to
have higher impact strength (5).

Crystallinity is never 100% complete; there are always some amorphous
areas (8). In contrast, certain polymers such as polycarbonates are
wholly amorphous. We will discuss crystallinity and amorphous structure and their effect on the fastening of plastics in more depth in
Chap. 3.

(a)
TYPICAL CHAIN
STRUCTURE

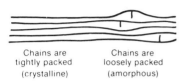

Chains are
tightly packed
(crystalline)

Chains are
loosely packed
(amorphous)

(b) HIGH-DENSITY POLYETHYLENE

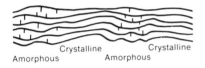

Crystalline   Crystalline
Amorphous   Amorphous

FIG. 1.3   (From Ref. 5.)

As indicated previously, the manufacturing process for thermoset plastics is quite different from that for thermoplastics, as the polymerization is done partly by the material supplier and finished in the molding process. Furthermore, thermoset chains cross-link in three dimensions, unlike thermoplastic polymer chains which usually grow lengthwise. "Rigid thermosets have short chains with many cross-links; (more) flexible thermosets have longer chains with fewer cross-links" (5).

Thermosets generally have more cross-linkage than thermoplastics, and this can result in improved dimensional stability and resistance to higher temperature. Application of heat to a cured thermoset will not cause it to remelt like a thermoplastic. However, too much heat applied to thermoset plastics will cause physical degradation and molecular breakdown.

Certain thermosets do not require added heat or pressure to polymerize. Polyurethanes, epoxies, and polyesters are ordinarily supplied in liquid, two-component form. Mixing the two components together causes an exothermic (heat-producing) cure or reaction. Because of the heat created, extra heat may not be applied, although it will speed up the cure.

In the future, distinctions between thermoplastics may become more blurred, as development of cross-linkage ability in thermoplastics becomes more widespread.

1.5 THERMOPLASTIC SUBGROUPS

Thermoplastic subgroups are determined by chemical structure, manufacturing method, feedstock (base raw material), and finished characteristics (such as texture and feel). They include olefins, vinyls, styrenics, and cellulosics. Certain thermoplastics are classified as "engineering grades," a bit of handy terminology invented to indicate that they have greater structural capacity.

1.5.1 Polyolefin Subgroups: Polypropylene and Polyethylene

These basic, relatively simple, structure thermoplastics offer low-to-moderate assembly strength with standard or special design thread-rolling screws for plastics. They are often suitable for use with specially designed push-in fasteners for plastics that are mechanically or ultrasonically installed with a straight-line, axial motion. Reasonably good performance can also be expected with ultrasonic installation of internally threaded metal inserts, except with ultrahigh-molecular-weight (UHMW) compositions.

1.5.2 Vinyls

The vinyl group is dominated by polyvinyl chloride (PVC). When unplasticized, in other words, when in their rigid form, PVCs offer fastening characteristics similar to polyolefins. Like polyolefins, PVC is often referred to as a commodity plastic owing to its widespread usage. The word "commodity" should not be considered a derogatory term. PVCs can provide a wide variety of assembly strengths.

1.5.3 Styrenics

This group includes acrylonitrile butadiene styrene (ABS), styrene acrylonitrile (SAN), polystyrene, and acrylics. They commonly provide high assembly strengths. In styrene-based polymers, we recommend the use of thread-rolling screws and push-in thread-forming fasteners, especially designed for plastics. Ultrasonic installation of internally threaded metal inserts is also recommended.

1.5.4 Fluoropolymers

Fluoropolymers such as tetrafluoroethylene (TFE) and fluoronated ethylene propylene (FEP) are very lubricious and have a "waxy" feel. To obtain reasonable torque levels only certain types of special-design screws for plastics should be used. Neither push-in-type thread-forming fasteners or ultrasonic installations are recommended.

1.5.5 Cellulosics

Cellulose is a polymer obtained from wood, cotton, soybean husks, and other natural fibrous materials. Cellulosics such as cellulose acetate and cellulose proprionate were early plastic developments. The materials are generally brittle, so that only thread-cutting fasteners and special designs of fasteners creating low-induced stress can be recommended. Ultrasonic installations will do well.

1.5.6 Engineering Grades

Polycarbonates, polyphenylene oxide (PPO), polysulfone, and some glass-filled olefins and nylons are examples of engineering-grade plastics. Plastics such as these are often chosen for their structural strength. Polycarbonates, modified phenylene oxides, and other engineering grades have also been developed in the form of structural foams having great strength.

For a variety of reasons, recommendation is generally for use of the special-design thread-rolling screws for plastics, sometimes for thread-cutting screws (nonfoam applications), or ultrasonic insertions.

1.5.7 Manufacturing of Structural Foams

The manufacturing process for structural foams involves the introduction of an inert gas, such as nitrogen, or premixed blowing agent into the resin. In one system, under low pressure, the gas expands as the material fills the partially filled mold. Another system uses high pressure and expandable molds or withdrawable inserts to accommodate expansion. Either method produces an internal cell-like structure with a tough, outer skin (Fig. 1.4). Special-design thread-rolling screws for

FIG. 1.4 *The Handbook of Engineering Structural Foam* is a useful source of information regarding structural foams made from engineering-grade plastics. (Courtesy of General Electric Company, Specialty Plastics Division, Pittsfield, MA.)

plastics perform well in structural foams. Thread-cutting screws should never be used in structural foams, as they will cut through the outer skin rather than compress it at the pilot hole entrance, thereby producing a weaker joint. Also, thread-cutting screws tend to tear away the internal foam without compacting it, rather than producing a smooth, stronger internal thread.

1.6 THERMOSET PLASTICS

Thermosetting compounds vary widely in their physical characteristics. They include phenolics, furan resins, aminoplastics, alkyds, allyls, epoxies, polyurethanes, polyesters, and silicones. All require review by a competent technical facility before threaded fasteners can be used. Ultrasonic insertion is not used with thermosets.

1.7 REACTION INJECTION MOLDING AND REINFORCED REACTION INJECTION MOLDING PROCESSES

Reaction injection molding (RIM) or reinforced reaction injection molding (RRIM) refer to processes for manufacturing two-component polymers that require little, or no, added heat to cure. The chief advantage of these processes is that they create little stress in the molded product.

RIM processes can be used to produce materials with very different characteristics, from very soft to stiff. We looked at some urethane RIM material which appeared to be fairly stiff and to have somewhat less cellular format than the structural foams made from polycarbonate. In our opinion, RIM urethanes at least lend themselves well to the special-design thread-rolling screws, but testing is recommended.

1.8 FILLERS

The use of glass fiber and mineral fillers can significantly enhance desired characteristics, such as strength and/or property balance, in thermoplastics. Such use has expanded the areas of application for plastics. It has also increased the need for structurally strong fastening sites, such as can be obtained by the use of thread-rolling screws, especially designed for use with plastics.

1.9 PLASTICIZERS

Plasticizers are added to some polymers, particularly PVC, to improve processibility and to reduce the secondary molecular forces holding the polymer chains together. This imparts flexibility. Addition of plasticizers has a significant effect on the use of threaded fasteners. (See Chap. 4, Sec. 4.3.1.)

Plasticizers work without destroying molecular structure. An illustration is the way water acts as a plasticizer when a housewife (or, in these days, perhaps husband) uses it to crease a shirt while ironing without destroying the material's fiber (9).

REFERENCES

1. R. E. Schulz, *Plastics Design Forum* 6:34 (March/April 1981).

2. The Plastics Production Process, *Plastics Design Forum*, Industry Media, Inc., Denver, CO.

3. *Plastics Design Forum* 6:35-38 (January/February 1981).

4. J. H. Crate, *Plastics Design Forum* 7:16 (March/April 1982).

5. *Machine Design*, 1982 materials reference issue, 54:96-98 (1982).

6. R. L. E. Brown, *Design and Manufacture of Plastic Parts*, Wiley, New York, 1980, p. 3.

7. E. Miller, *Plastics Products Design Handbook*, part A, Marcel Dekker, New York, 1981, p. 8.

8. L. Berringer, in the lecture series *The Plastic Seminars*.

9. F. Rodriguez, *Principles of Polymer Science*, McGraw-Hill, New York, 1970, p. 46.

BIBLIOGRAPHY

Agranoff, J., Ed., *Modern Plastics Encyclopedia*, Vol. 58, No. 10A, McGraw-Hill, New York,

The Handbook of Engineering Structural Foams, General Electric, SFR-3, January 1978.

Machine Design, 1982 materials reference issue, 54 (1982).

Plastics Design Forum, focus issue (April 1981).

2
MECHANICAL FASTENERS USED IN PLASTICS

2.1 PERMANENT ASSEMBLIES

The present discussion will not dwell on product assemblies designed to be permanently unitized or connected. In these cases, designers have a wealth of joining methods at their disposal, including space-age adhesives; hot melts; solvent bonds; friction heat, ultrasonic, induction, dielectric, hot platen, and hot gas welding.

Various types of rivets are also widely used in permanent assemblies, although they generally provide lower tensile and fatigue strengths than comparable screws do. Pins of various types are effective and inexpensive where loading is primarily on shear and prevailing torque is not a prime factor.

2.2 THE CHALLENGE: DISASSEMBLY

Parts which must be disassembled for servicing or replacement present the ultimate test of the fastening method employed. In some cases, the use of an actual mechanical fastener can be avoided by designing hinges, latches, or detents into the assembly. However, these methods can only be used with plastics capable of withstanding the strain of assembly, loads in service, and repeated disassembly. They are recommended only for lightly loaded, nonrigid product design.

For the most part, mechanical fasteners must be used for nonpermanent assemblies. These include a broad range of threaded and unthreaded devices, both metal and plastic. Some were originally designed for use in metals; others are specifically intended for plastics applications.

Material in this chapter has been both adapted and directly extracted from *Assembly Engineering*, August and September, 1982. By permission of the publisher © 1982 Hitchcock Publishing Co. All rights reserved.

2.2.1 Nonthreaded Fasteners

Press-in-type fasteners or lock nuts and clips are sometimes used where thin sections of plastic must be joined. They can be either forced, glued, or expanded into molded or drilled holes; molded in place; or inserted ultrasonically. Of the press-in fasteners, we will examine only the threaded, press-in fasteners which are defined as "push-in thread-forming fasteners for plastics."

2.2.2 Nuts and Bolts

Machine nuts and screws, seldom used in plastics by themselves, have inherent disadvantages. Later in this chapter, we discuss the creation of stress from use of machine screw threads in standard thread-forming screws. The tightening torque required with use of machine nuts and bolts per se, without any compensation, could damage the plastic, with cracking frequently occurring under torque loads due to inherent stresses in the plastic. Also, owing to the creep of some plastics, clamp loads often tend to fall off over a period of time with resultant loosening of the nut.

If nuts and bolts are used, consideration should be given to the use of washers to spread out the created tightening stress, as far as possible. Machine screws are successfully used, however, in plastic applications with internally threaded metal inserts.

2.2.3 Inserts

Special forms of nuts, called inserts (Fig. 2.1), are frequently used to provide threaded holes in plastics, in both open and blind locations. They include nonthreaded types for lightly loaded applications as well as self-tapping, solid-bushing, molded-in, and ultrasonically inserted versions. As mentioned previously, inserts are generally for use with standard machine screws.

 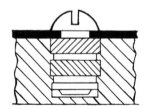

FIG. 2.1   Insert with machine screw. Installation may be made into thermoplastics either ultrasonically or by heat methods. Such inserts can be used successfully in polycarbonates, modified polyphenylene oxide, and other of the newer engineering grade polymers. (Courtesy of P.S.M. Fasteners, Ltd., Willenhall, West Midlands, England.)

FIG. 2.2   Twin-tite® stud for plastics.

This type of fastening should be reserved for applications where very frequent disassembly is required, since the use of the insert represents extra cost. There may be exceptions owing to other factors.

2.2.4   Studs

Double-ended, threaded studs (Fig. 2.2) can also be used for frequent disassembly. In this case, the exterior end of the stud is also provided with a machine screw thread. The other end can have a thread especially designed for plastics. The washer, in the middle of the stud, should be large enough to dissipate stress created in tightening the fastening.

2.2.5   Plastic Fasteners for Plastics

Fasteners made of plastic are available in both threaded and unthreaded versions, generally for specialized usage. They include bolts, screws, rods, studs, and nuts. Unthreaded types include inserts, clips, clamps, and special quick-operating fasteners. Threaded types have the same standards for dimensions, thread class, and fit as do threaded machine screws and nuts.

Use of such fasteners can be necessary under certain corrosive conditions, in which metals would be attacked. They are nonmagnetic and have the chemical resistance of the material from which they are manufactured. When they are used, consideration should be given to any physical differences (for instance, in expansion rate) between the material used for the screw and the material from which the nut member may be made.

At the present time, fasteners manufactured from plastics reportedly are available in nylon (6/6), (rigid) polyvinyl chloride (PVC), acrylic, acetyl, polyethylene, polypropylene, TFE fluorocarbons, and polyimide. There may be other materials used for fasteners of which we are currently unaware.

Screws molded from plastic materials do not generally provide sufficient hardness or strength for use in forming threads, although plastic screws with machine screw threads are available with steel cores. This provides added strength, but at the sacrifice of some of the abilities mentioned in the preceding paragraph.

In this field there always seem to be exceptions, and in the case of threading ability we have one particularly interesting exception. Screws have been developed from Vespel polyimide, which reportedly can form their own threads in pilot holes in an aluminum. Use is highly specialized of course, and a combination of insulation characteristics and high-temperature ability of the polymer is required. Compared to screws manufactured from metal the cost is very high; but it is obvious that cost in such specialized usage is not the primary criterion.

2.2.6 Machine Screws Used in Tapped Holes

If metal or plastic machine screws are used in a tapped hole in plastics, tap selection and usage should be done in order to avoid creating excess heat in the non-heat-conductive plastic nut member. To help avoid excessive heat creation, tap manufacturers recommend narrow lands on taps used in plastics. This may call for using an odd number of flutes (say, five or three). Selection of tap size and fit will have to take into consideration the ductility and elasticity of the plastic which will tend to cause a partial shrinkage. Further remarks on tapping in plastics will be found in Chap. 6.

Molded-in male and female threads are used mostly for pipes and valves. They are not often used in the kinds of designed applications we are chiefly concerned about in this volume. Molded threads increase the molding cost significantly owing to added mold complexity and added time spent in handling threaded core pins.

2.3 METAL THREAD-FORMING AND THREAD-CUTTING SCREWS FOR PLASTICS

The history of the development of fasteners has always been closely related to the development or invention of new materials. The debut of any kind of fastener naturally follows the invention of the material to be used in the manufacture of various products. This has been so ever since man first conceived the idea of fastening a couple of logs together with vines to make it easier to paddle across rivers. The screws used in fastening plastics follow this natural progression.

First came the development and substitution of plastics for other materials. But when product designers attempted to use existing threaded fasteners with these new materials, difficulties began. Often these difficulties stemmed from the internal stresses within the product which were created by the plastic-manufacturing process. Introduction of screws into the already stressed material exacerbated the problem.

In order to combat the introduction of added stresses when inserting screws, fastener manufacturers have recently developed screws that are especially designed for use in plastics. Some of these also take advantage of inherent characteristics of the plastic to improve joint strength. However, before getting into the subject of special screws, we will discuss standard screws used in plastics.

2.3.1 First Screws Designed for Metals

The first self-threading screws which were tried in plastics were the standard, round-body types AB, B, and C sheet metal screws used for some 60 years or more in metals. These screws are known in the fastener industry as thread-forming screws because they form or tap their own threads in drilled or cored holes—without the need to pre-tap. And because they were developed for use in metal, they are case-hardened and stress-relieved.

We shall try to always refer to these screws as "standard thread-forming screws." They are shown in Fig. 2.3. Most of us are familiar with them. When plastics first came under consideration for structural applications, these were the only fasteners available. By specifically calling these screws "standard thread-forming screws" we differentiate them from the newly developed and specially designed screws for plastic which are discussed later in this chapter.

Figure 2.3 shows the type C thread-forming screw which has a 60° machine screw thread. The use of thread-forming screws *in metal*, whether they be standard or specially made to provide other benefits, generally tends to provide a strengthened joint. The thread produced in the forming action has been pressed into the inside of the pilot hole. With metals, the grain flow of the nut member using type C screws would follow the path illustrated in Fig. 2.4.

Figure 2.4 shows the expected strengthened nut member threads produced by such a thread-forming fastener. However, in plastics the results can be different. The machine screw threads are too close together. When plastics are compressed, very often cracks are created at notch-sensitive areas at the root of the plastic's internal thread. Consequently, standard type C thread-forming screws are not often used in plastics.

After considering the stress created by type C standard thread formers (Fig. 2.4), designers attempted to use the type AB screw because of its wider thread spacing. There should be expected benefits, and

there are. But the gimlet point on a type AB screw requires extra-
hole depth to provide sufficient length of thread engagement. So they
turned to specifying type B (Figs. 2.3 and 2.4), certainly an improve-
ment because of the blunter point design. It is the best *standard*
thread-forming screw to use in most applications, and we will return
to it later.

Type	ANSI Standard	Manufacturer
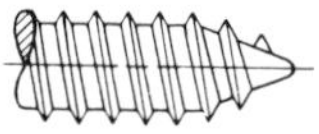	AB	AB
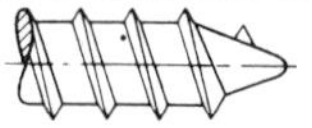 NOT RECOMMENDED—USE TYPE AB**	A	A
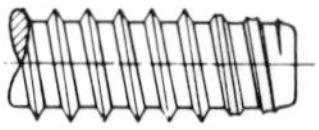	B	B
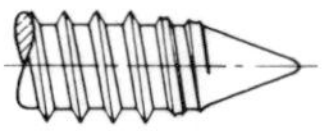	BP	BP
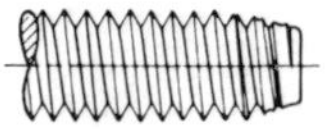	C	C
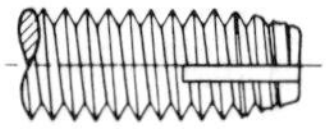	D	1
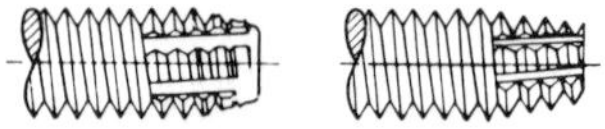	F	F
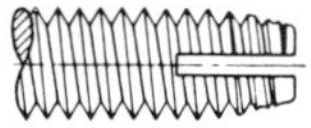	G	G
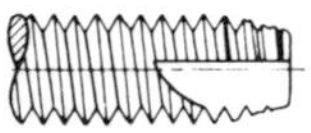	T	23
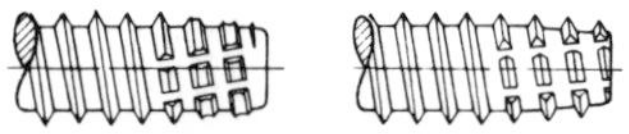	BF	BF
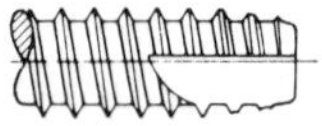	BT	25

FIG. 2.3   Standard thread-forming screws. (Courtesy of The American Society of Mechanical Engineers, New York.)

Plastic Nut Member
Simulated Plastic Displacement-Compression Pattern

Crack Caused by Overcompressing the Plastic

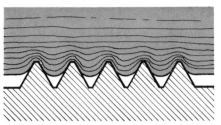

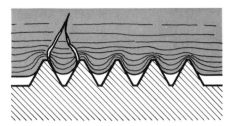

Type C Thread Forming Screw
Compression pattern that might be expected

Type C Thread-Forming Screw
Plastics do not have the cold-flow characteristics of
steel and cracks can be created at notch sensitive
areas at the thread crests

Type B Thread Forming Screw

Plastic Nut Member
Crack Caused by Overcompressing the Plastic

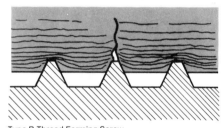

Wider thread configuration of Type B thread former

Type B Thread-Forming Screw
Even with the Type B thread there can be
overcompression accompanied by cracking at notch
sensitive locations

FIG. 2.4   (From *Assembly Engineering*, Sept. 1982. © 1982, Hitchcock Publishing Co., Wheaton, IL.)

Both types AB and B screws achieve wider thread spacing than the
standard machine screw by merely eliminating every other thread, for
example, a #10-32 machine screw versus a #10-16 tapping screw.

All three standard types AB, B, and C thread-forming screws have
been successfully used in the rather ductile commodity plastics such
as polypropylene and polyethylene, when unmodified by the addition
of fillers or in low modification. However, in other more structural
plastics the relatively high torque required to insert any one of these
three standard thread-forming screws tends to further compress the
plastic and may result in cracking of the material at the fastener site
(Fig. 2.4). To avoid this difficulty, these types are usually recommended for insertion with a shallow thread.

Shallow threads do not allow full fastening strength to be developed at the fastening site.

2.3.2 Qualified Success Achieved

Of the three types of standard thread-forming screws, type B appears to be preferable owing to its wider-spaced thread, and because it does not have the space requirement of the gimlet point found in the type AB screw. This wider-spaced thread creates less stress, and the blunt point permits a greater engagement length.

Standard Thread-Cutting Screws

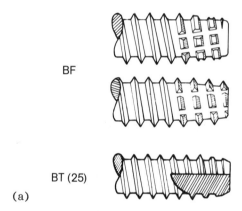

BF

BT (25)

(a)

Plastic Nut Member
Section of Thread Formed by Type BT TCS

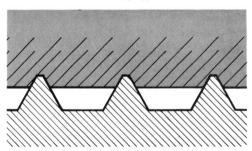

Type BT Thread-Cutting Screw
The shallow thread with thread-cutting screws means
(b)   low pullout strength

FIG. 2.5 (From *Assembly Engineering*, Aug. 1982. © 1982, Hitch-cock Publishing Co., Wheaton, IL.)

2.3.3 Thread-Cutting Screws

Also listed among the standard self-threading products are thread-cutting screws. These are available with a variety of cutting points. The best performers in plastics are the type BT or 25 screws (Fig. 2.5). These have a wide-spaced thread and a slash-type cutting edge.

Another type of thread-cutting screw is the type BF. It is cheaper and simpler to manufacture because the cutting flutes are introduced in the manufacturer's thread-rolling process rather than in a separate operation. Unfortunately, the BF thread tends to cut a rather messy thread in some ductile plastics because of chip interference and flute clogging.

At least one prominent fastener manufacturer has recently introduced a special-design, thread-cutting screw similar to the standard BF. It is called the BF Plus*. Thread spacing on the BF Plus is wider and the cutting edges are reported to be sharper than on the standard product.

The two, somewhat abbreviated laboratory tests which we have seen on the BF Plus screw were conducted in phenolic and contrasted this thread-cutting screw with standard, type BT (25) thread cutters. Results appear to give approximately the same, or slightly higher differential than the standard BT product in this difficult-to-thread material.

Although there are drawbacks, thread-cutting screws of the BT type are often recommended for use in plastics. On the plus side, they do not create much stress when used in the recommended hole sizes, however, this standard pilot hole size recommendation also results in a shallow thread (Fig. 2.5). In addition to providing stress relief, the shallow thread is partially required to keep installation torque down to a reasonable level. A manufacturing problem with thread-cutting screws which may lead to shallow threading lies in the manufacturing tolerances that are required.

The point design of ordinary, standard thread-cutting screws does not have the rake and clearance angles that are found on more efficiently performing ground thread, thread-cutting taps. In screw production, the secondary manufacturing operation of point-slotting, as done under current fastener manufacturing practice, is not capable of consistently producing the same quality cutting action as found with a high speed ground thread tap.

Plating and bulk-handling methods used with screws further reduce cutting efficiency by dulling the cutting edges.

In the case of blind holes, the slashed flute of the BT thread-cutting screw requires a deeper hole to allow for chip production, which can be a design inconvenience. [Approximately 20% of nominal screw diameter should be allowed for chip clearance (1).] *More importantly,*

---

*BF Plus is a tradename of Camcar/Textron, Rockford, Illinois.

since the flute takes up several threads, considerably less holding power is available from the long tapered threads at the cutting point. As a result, full fastening strength cannot be achieved with thread-cutting screws unless working with a through hole—as they were intended to be used, when developed for use in metals.

The assembly-line operator should be advised to start thread cutters rather carefully, since this screw tends to continue to follow whatever direction it is started in. If inserted at an off-angle, in plastic there will be strippage or cracking of the boss.

Thread-cutting screws are often not chosen for use in electronic or electrical applications as the chips or debris from cutting could mess up the assembly.

In spite of the drawbacks there are materials for which no screws other than thread cutters are currently available. For the most part, they are used with certain thermoset plastics and sometimes with certain engineering resins, as shown in the summary of screw usage in plastics in Appendix 3. However, it is our opinion that owing to the generally shallow thread depth, the decision to use them will defeat the achievement of full fastening strength.

Regarding types D and F thread-cutting screws (Fig. 2.3), their flutes have proved ineffective for plastics. Their use is limited.

2.4 NEED FOR NEW SCREW TYPES

As the market and use of structural plastics spread, and new types of stronger plastics developed, it became apparent that new fasteners were required. As a result, within the last 5 to 10 years, threaded-fastener manufacturers have met the challenge and provided improved screws, especially designed for use in the new materials.

These fasteners for plastics were designed in order to overcome the objectionable features of the standard sheet metal screws when used in plastics—such drawbacks as high drive torque and induced stress, also shallow thread and low pullout which tend to go hand in glove. (Pullout is an indication of joint strength.)

2.4.1 High Drive Torque

High drive torque with standard, round-body, thread-forming screws results from the friction created by a 360° circumference in contact with the plastic. (Friction is the highest ingredient in fastener-drive torque.) Actual drive torques tend to vary erratically on the assembly line.

High drive torque can result in difficulty with adjustment of driving tools if not accompanied by a change in fail torque. This difficulty is due to a restriction in the differential between fail and drive torques. Furthermore, a high drive torque in plastic assemblies results in less

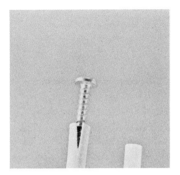

FIG. 2.6   Radial stress causes cracking of boss.

available tightening torque. This occurs because of intrinsic stress within the plastic—a limiting factor on the amount of tightening torque that can be applied.

2.4.2 Induced Stress

The second problem is that standard thread-forming screws create stress. The specific name for this inherent problem in plastics is *hoop stress*. In this condition, stresses are radially compounded in the plastic around the outside chamber of the screw in a hoop configuration. They tend to crack bosses (Fig. 2.6) and prevent effective threading. Thus, with the ordinary self-threader only a shallow thread can be made owing to the necessity of keeping drive torque at a minimum. Also, of course with a shallow thread you would expect to get low pullout and less joint strength, and you do.

2.5 THREAD-ROLLING SCREWS ESPECIALLY DESIGNED FOR USE IN PLASTICS

To avoid the problems of high drive torque, induced stress, and pullout, farsighted fastener manufacturers developed new screws, especially made for use in structural plastics fastening. About 20 years ago, when fastener manufacturers first began to consider using screws in plastics, they tended to stay with readily obtainable tooling. The initial product developments had a 60° thread and could use the existing thread-rolling dies that were inexpensive and quick to obtain, but with the development of stronger, harder, and more structural plastics, the manufacturers had to have second thoughts.

2.5.1 Characteristics of the New Screws for Plastics

The common characteristics of the more successful screws that have recently been expressly invented for the new materials are as follows:

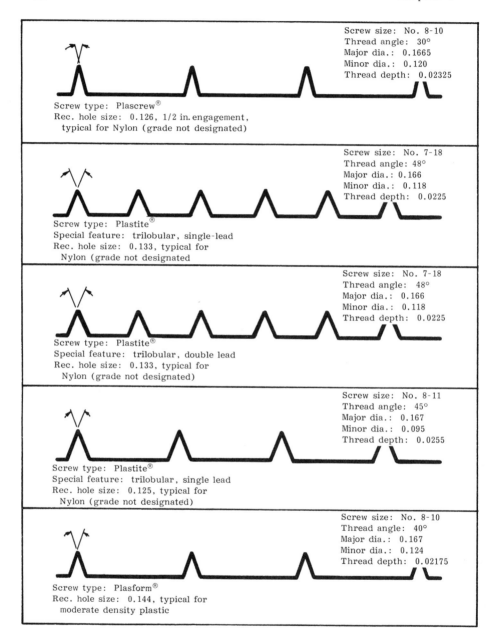

FIG. 2.7   Examples of special-design thread form used in plastics. (From manufacturers' catalogs. A list of manufacturers of these screws will be found in Appendix 1 under the Trademark section.)

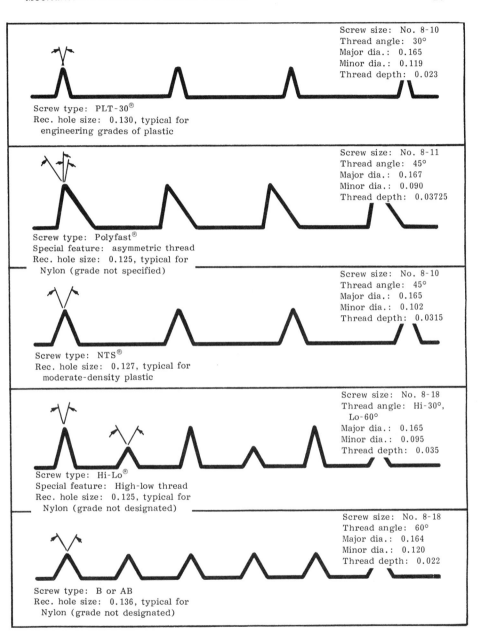

Screw size: No. 8-10
Thread angle: 30°
Major dia.: 0.165
Minor dia.: 0.119
Thread depth: 0.023

Screw type: PLT-30®
Rec. hole size: 0.130, typical for
engineering grades of plastic

Screw size: No. 8-11
Thread angle: 45°
Major dia.: 0.167
Minor dia.: 0.090
Thread depth: 0.03725

Screw type: Polyfast®
Special feature: asymmetric thread
Rec. hole size: 0.125, typical for
Nylon (grade not specified)

Screw size: No. 8-10
Thread angle: 45°
Major dia.: 0.165
Minor dia.: 0.102
Thread depth: 0.0315

Screw type: NTS®
Rec. hole size: 0.127, typical for
moderate-density plastic

Screw size: No. 8-18
Thread angle: Hi-30°,
  Lo-60°
Major dia.: 0.165
Minor dia.: 0.095
Thread depth: 0.035

Screw type: Hi-Lo®
Special feature: High-low thread
Rec. hole size: 0.125, typical for
Nylon (grade not designated)

Screw size: No. 8-18
Thread angle: 60°
Major dia.: 0.164
Minor dia.: 0.120
Thread depth: 0.022

Screw type: B or AB
Rec. hole size: 0.136, typical for
Nylon (grade not designated)

FIG. 2.7 (Continued)

1. A wider-spaced thread than that provided by the old, standard
    thread formers (types AB or B tapping screws) which were orig-
    inally designed for use in metals.
2. An internal thread angle which is typically less than the usual
    60° angle, long used on screws intended for metals. The in-
    cluded thread angles of the new screws vary from a low of about
    30° up to 48°.

Examples of screw threads that conform to these design features are
shown in Fig. 2.7.

The reasons for the difference in screw design lie in the disparity of
strength and susceptibility to cracking as between metals and the new
plastics when they are used as nut members. The extra-wide thread
spacing and the more acute thread angle of these new designs keep induced stress to a minimum during the thread-forming operation. They
also permit a deeper thread without increased drive torque.

For screws used in metals, holding power is largely determined by
the number of threads that effectively bite into the wall of the hole.
Usually with metals, more total fastening strength is obtained from the
number of threads being engaged than from depth of thread.

The extra-wide-spaced thread of the special-design screws for plastics allow stress to be dispersed over a greater area of the nut member,
and the plastic material remaining between the threads is left strong
and intact. On the other hand, if finer-pitch screws are used in the new
plastics, the plastic nut member material between the closely spaced
threads becomes overly stressed, is more susceptible to cracking, and
will not support higher loads or provide stripping resistance.

The narrower thread angle also reduces stress by displacing less
material and creating less bursting pressure (Fig. 2.8).

Of course, the material of the hardened and annealed metal screws is
significantly stronger than the plastic nut member material. Therefore,
if a plastic assembly is overloaded, the plastic will give way and strip
out. Stripout in the plastic will occur cylindrically and the more nut
member material that remains within the cylinder, the more resistant
to stripout will be the plastic. Therefore, in plastic assemblies using
a metal screw, wide spacing of threads increases assembly strength at
the same time that it reduces stress.

2.5.2   Double-Lead Threads

In attacking the problem of providing increased differential between
drive and fail torque, fastener designers have incorporated other features as well, such as the use of double-lead threads. The steeper
helix angle from the double lead permits a higher strip or fail torque
to be obtained, and usually a wider differential between them which
allows easier power tool adjustment. A double-lead fastener also enters

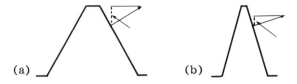

FIG. 2.8 (a) Standard thread profile. Bursting pressure. (b) Narrow thread profile. Bursting pressure.

the hole faster, and the site design must accommodate the higher speed and possible resulting higher impact.

2.5.3 Trilobular Shape

An additional approach (and one that has been widely licensed for manufacture all over the world) is the use of a trilobular shape on the screw shank. The trilobular design provides decreased drive torque (Fig. 2.9) and increased resistance to stripping (higher available strip torque) and increased differential. It also makes use of the natural resiliency of plastics to help lock the fastener in place, providing prevailing locking torque, by allowing the plastic to fill in the depressions between the lobes.

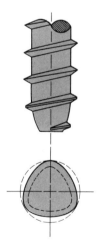

FIG. 2.9 Plastite 45 trilobular thread-rolling screws have extra-wide spacing between narrow profile threads resulting in exceptional low-induced stress and extra-low-boss-bursting tendencies.

2.5.4   Small Root Diameter

To give extra clearance for displaced plastic material, thereby further
reducing induced stress with thin-wall bosses, screw manufacturers
sometimes make their products with quite narrow root or minor diameters (Fig. 2.10). Subject to laboratory testing for the particular
application, this may prove out well. In the screw manufacturing process, it is sometimes difficult with certain thread designs to entirely
roll a smooth minor diameter. In general, this should not affect performance.

2.5.5   Semibuttress Thread Form

This feature used by one manufacturer appears to be intended to improve pullout. After examining test results, our opinion is that when
helically installed, the semibuttress thread does not appear to offer
any significant advantage over the other special design screws for
plastic, although the design may favor ease of alignment.

2.5.6   "Push-in" Fasteners

Another kind of semibuttress, helically threaded fastener for use in
ductile, resilient plastics is the press-in or push-in type (Fig. 2.11).
These are designed for straight-line, axial-load equipment or for ultrasonic assembly. With the helical thread provided, once the fastener is
installed, a mating thread is formed by the plastic. They can be rotationally removed and reinserted. The strength of an assembly utilizing
push-in fasteners can be quite surprising.

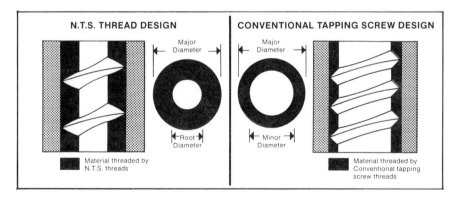

FIG. 2.10   Comparison of material held related to thread surface.
(Courtesy of Reed and Prince Mfg. Co., Jaffrey, NH, and Worcester,
MA.)

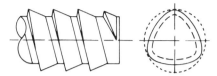

FIG. 2.11 Pushtite® II thread-forming fasteners are designed to be pressed into pilot holes with a single, straight-line stroke. They have a trilobular thread form. Threads are helical, not annular, which allows the fasteners to be removed and reinserted. (From *Assembly Engineering*, Aug. 1982. © 1982, Hitchcock Publishing Co., Wheaton, IL.)

Push-in-type fasteners are recommended for ductile plastics because they take advantage of the resilient characteristics of these materials which form back around the thread. Their advantages are economy of assembly and elimination of torque-related assembly problems. The fasteners can be made tamper-resistant by elimination of the driving feature. When installed with conventional press-type equipment, push-in fasteners should be reserved for nonstructural applications such as audio tape decks or even spatula handles. However, they can also be installed ultrasonically for more structural-type applications.

Manufacturers of helically threaded push-in fasteners are listed in Appendix 1.

2.7 ONLY GUIDANCE CAN BE GIVEN

Specific recommendations for the use of specialized and standard self-threading fasteners in all the various designs and in all the various available material compositions cannot be given in detail. Even in this book covering most of the existing generic plastics with which screws are used (Chap. 4) and many tradename plastics (Chap. 5), there are just too many variations. However, we are of the opinion that in most cases performance advantages are obtained using the fasteners specifically designed for use in plastics. Standard thread formers are best confined to use in the ductile, commodity plastics and in nondemanding applications. We also very strongly recommend that a fastener laboratory be consulted for evaluation of all new applications. Chapter 8 discusses the fastener laboratory.

REFERENCE

1. R. L. Reynolds, *Assembly Engineering* 14:27 (September 1971).

BIBLIOGRAPHY

Gomes, K. J., *Assembly Engineering* 25:24-27 (August 1982).

Reynolds, R. L., *Assembly Engineering* 14:26-30 (September 1971).

Wagner, D., *Assembly Engineering* 23:26-29 (November 1980).

3
FASTENING CHARACTERISTICS
OF PLASTICS

This chapter will provide a modest contribution to our understanding of the factors that influence the behavior of plastics, so that when we come across an unfamiliar material we will know what to look for to pre-assess its fastening characteristics. Unfortunately, in a volume of this size there is no way that we can cover all the possible materials, combinations of materials, grades, or variations in filled content that even now exist.

Before surveying the use of the threaded fasteners in the specific plastics that we do cover generically (see Chap. 4) and by tradename (see Chap. 5), some general characteristics of plastic materials can be considered—universal characteristics that fall under the broad heading of material stability, namely, stiffness and elasticity. With regard to stiffness we will discuss flexural modulus as a rough measure of stiffness and the effects on stiffness of reinforcements, crystallinity, molecular weight, and moisture content. Under the heading of elasticity we will briefly cover creep, cold flow and resiliency, and material relaxation.

3.1 STIFFNESS IN PLASTICS

The degree of stiffness of a material has profound consequences on the ability of a plastic to accept metal fasteners without cracking, on the torque levels obtained, and on the general strength of the fastening site.

3.1.1 Flexural Modulus: A Measure of Stiffness

Flexural modulus is a measure of a material's degree of stiffness; it has been called the modulus or coefficient of bending, as expressed in pounds per square inch (psi). The higher the degree of stiffness or rigidity, the greater the flex modulus. Minerals and other fillers and fibers (principally glass), when blended with the polymer, add to the stiffness. It will be seen that the flex modulus can be of modest assistance in a generalized appraisal of threading in plastics. Chapter 4 includes the corresponding modulus for each polymer discussed.

By grouping plastics into four tentative, flex moduli ranges, it is possible to make some general observations about the use of standard thread-forming screws and specially designed thread-rolling screws—subject to specific material review of all the characteristics that may affect threading. In general, the higher the flex modulus value the greater is the need for the more sophisticated fasteners. For self-threading screws and inserts installed in the same manner, the higher the flex modulus the higher the required drive torque, fail torque, and pullout tensile values will be.

3.1.2 Polymers with Flex Moduli Less Than 200 × 10³ PSI

In highly ductile materials having flexural moduli under 200,000 psi (with the exception of fluoropolymers), standard thread-forming screws give generally satisfactory performance. Of course, we believe much better performance will be achieved, when it is required, with the special-design thread-rolling screws for plastics. We also recommend that the push-in thread-forming fastener be considered for nonstructural usage in these materials. Such fasteners are available with helical threads and either round- or trilobular-body shapes.

The considerable, elastic memories of highly ductile plastics cause the material to close in around the thread form of the fastener after insertion and provide adequate holding power.

The major manufacturers of the push-in-type fastener with helical threads are listed in Appendix 1.

Standard thread-cutting screws tend to cut rather poor quality threads in these elastic plastics, but can be used in nondemanding applications, where the chances of screw removal and reinsertion are minimal.

3.1.3 Polymers with Flex Moduli from 200 to 400 × 10³ PSI

With polymers having flex moduli within the 200,000—400,000 psi range, standard thread formers are often avoided because of induced stress in the fastened part. Our recommendation is to turn to the special-design thread rollers for plastics (if the application is more than positional), otherwise to consider push-in thread-forming fasteners. For applications which will require multiple reassemblies, ultrasonic installations of internally threaded inserts for use with standard machine screws generally give excellent results.

As they age, plastics containing plasticizers such as acrylonitrile-butadiene-styrene (ABS) and polyvinyl chloride lose elasticity. They become more brittle because of evaporation of the plasticizer (1). Owing to this factor, reinsertion of screws must be done at lower torques. When a fastener is reinserted, a lower amount of applied torque will be used up in forming or rolling the thread and a greater amount will be available for compressing the load. Higher compression will result in more brittleness and more opportunity for the joint to fail.

3.1.4 Polymers with Flex Moduli from 400 to 1000 × 10³ PSI

Many of the materials having flex moduli in the 400,000–1,000,000 psi range, achieve their degree of stiffness through the addition of glass fiber or other filler. Special thread-rolling screws or thread-cutting screws are usually the best performers in these materials. Details will be discussed in Chap. 4.

In most cases, by virtue of creating less stress, the special-design thread-rolling screws for plastic will provide the most satisfactory performance. Thread-cutting screws will cut through the fibers which arrange themselves around the cored pilot hole during the molding process. This condition will result in a weaker joint.

Our specific recommendation among the special-design screws for plastics would be for those having thread angles between 40 and 48 degrees and with the widest pitches, consistent with required engagement length.

3.1.5 Polymers with Flex Moduli Over 1000 × 10³ PSI

Materials having flex moduli above 1,000,000 psi can be divided into two types: (1) thermoplastics that achieve such a degree of stiffness from added reinforcement and (2) the thermosets that have such stiffness without reinforcement. With all materials of either kind we strongly urge that consultation be made with a competent fastener laboratory or the resin supplier.

When working with thermosets, if many reassemblies are to be made, the fastener specifier should consider using internally threaded metal inserts *with exterior, coarse cutting threads or expansion-type inserts.*

Most thermoset plastics, such as phenolics, tend to be very brittle and granulate in between the fastener threads (1) (Fig. 3.1). Thread-cutting screws, which are usually the screws recommended, should be of the type T (25) design. They also have coarse pitch threads (Fig. 2.3).

Many fastener customers specify coarse thread pitches in phenolics when using special design thread-rolling screws. Thread-rolling screws are not recommended for much more than one insertion. We know of two

FIG. 3.1   Friable plastic materials crumble so easily that just removing the screw sometimes destroys the mating thread. (From *Appliance Manufacturer*, © March 1979, Cahners Publishing Co., Inc., Boston, MA.)

manufacturers of the special-design screws for plastics that also make them with cutting points specifically for use with this stiff-type material. Their addresses are listed in Appendix 1 in the section covering special-design screws for plastic.

With some grades of thermoset polyesters, we strongly urge inquiry into the special-design, coarse thread rollers for plastics. Laboratory tests show many excellent results in thermoset polyesters (high torque, less stress, and reusability).

3.2   REINFORCEMENTS: EFFECTS ON PLASTICS

Reinforcement of plastics by the addition of either powdered or granulated minerals or by the addition of glass (or other) fiber is an extremely common practice. The purpose is twofold; (1) the use of fillers and reinforcements can upgrade the mechanical characteristics significantly; and (2) the reinforcement can also reduce the overall cost of material in the finished product.

Far from merely being material extenders, minerals such as talc, silica, mica, calcium carbonate, and possibly others can change the material properties of a polymeric blend in a variety of advantageous ways—when properly mixed and combined with coupling agents. On the other hand, you might not want to blend extremely fine substances into compositions where ultrasonic insertion was contemplated, as the superfine particles could affect melt viscosity. Mineral mixed in with the polymer can add rigidity and stiffness to the cured product. If so, the creep will be reduced. Fibers up to about 1/2 in. will also add strength to the processed part and reduce the creep.

The fiber we are presently and most generally concerned with in commercial practice is composed of glass filaments, although carbon and even boron filament material can be used (currently at very high prices) for highly increased mechanical and physical abilities. Coupling agents are used to promote the bond between the glass and the polymer.

Fiber length and degree of perfection of the glass filaments can also affect the composite's strength. The mechanical characteristics of the composite material will be affected by the way any long fibers are oriented in the mold. If long fiber is used and is oriented in one direction only, the resultant product will be stronger and stiffer in one direction only. This may or may not be desirable.

In general, whenever possible, we would prefer to work with thread-rolling screws in glass fiber composites since use of thread-cutting screws will reduce joint strength by cutting through the fibers at the joint. Even though the fibers may be short (say, 1/8 in.), the "trade-off" in using glass fiber is that unless it is completely immersed in the polymer the ends of the filaments will suck up moisture and reduce the strength of the glass-to-polymer bond. This is a factor to be considered in some cases when using thread-cutting screws.

We have known extremely successful use of thread-rolling screws with up to 30% glass fiber content by volume in the engineering grades of materials. Beyond that, we are unable to comment except to recommend that applications be tested at a fastener laboratory. With ultrasonic insertions, 40% glass fiber content appears to be about the limit. When glass fiber content exceeds this amount (and we understand it is available with some materials, such as Nylon), then examination should be made of thread-cutting fasteners.

In general, the higher the glass fiber content, the greater the required torque to form threads with self-threading fasteners.

Additional information on reinforcement in the more common plastics will be found in Chap. 4. A recommended initial source for general information on the subject of fillers, extenders, fibrous reinforcements, and indeed the whole subject of polymers is *Modern Plastics Encyclopedia* published by McGraw-Hill.

3.3 CRYSTALLINITY IN PLASTICS

In Chap. 1, Sec. 1.4, we referred briefly to the effects of crystallinity on a plastic's rigidity and strength; we now discuss this in detail.

Although it does not appear to be complete, considerable research has been done in exploring crystallinity. It appears that there is an order to the structure of crystalline polymers. Among other researchers, E. H. Andrews and co-workers (2) explored the subject in some degree at Queen Mary College, London. Their report is an explanation—in part—of the effect on mechanical properties by the amount and orientation of the crystalline segment of a polymer's microstructure and the orientation and state of the amorphous matrix. Andrews indicates that crystalline-amorphous structure is controlled not only by the microstructure of the polymer but also by the physical conditions of temperature, time, and strain of solidification (2).

R. L. E. Brown, in his clearly written volume (3), summarizes the subject of crystallinity as follows:

> Crystalline polymers are essentially two-phase systems, with a crystalline phase . . . consisting of regions of ordered, regular molecular arrangements . . . dispersed in . . . a disordered . . . amorphous matrix. The . . . crystalline regions can be large.

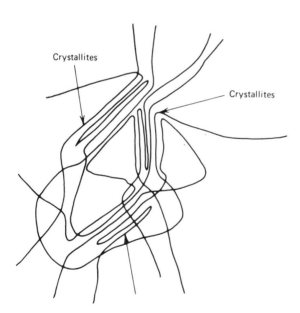

FIG. 3.2 Polymeric chains spontaneously fold after achieving a certain length. The folded-chain structure permits closer packing and a higher degree of crystallinity. (From Ref. 3, p. 8.)

Figure 3.2 is a representation of a possible arrangement of molecules in a crystalline polymer.

Polyethylene, for example, can be obtained in a form in which 95% of its volume is crystalline in arrangement.

The effects of crystallinity on the properties of polymeric materials can most easily be seen by comparing the different forms of polymer that can exist within the same generic family. The highly crystalline form of polyethylene mentioned earlier is known as high-density polyethylene (specific gravity 0.97). Polyethylene also exists in low- and medium-density forms. The low density form is approximately 60% crystalline (specific gravity 0.915).

The high-density polyethylene is stronger, tougher and optically clearer than the other forms.

In the case of polyethylene, the difference in density arises mainly from the existence of short (two to six carbon-atom length) branch chains in the low-density material, resulting from side reactions during polymerization. These branched polymer molecules do not as easily pack together in an orderly fashion, and thus lower the crystallinity and the density of the material. Also, since the polymer chains are more closely packed in the crystalline areas than in the amorphous areas, they are therefore in more regular and close contact over relatively long distance.

The net secondary forces holding the polymer chains in position one with respect to another in the crystallites are therefore greater than in the amorphous region, with the result that increasing the crystallinity tends to increase the strength and rigidity of the plastic. (3)

The degree of crystallinity while it is usually measured by x-ray diffraction techniques, and more sensitively by certain mechanical tests (4), can be estimated by reference to the specific gravity. The effects of higher crystallinity are as follows:

1. Increased stiffness, rigidity and resistance to creep, and increased threading torque.
2. Higher tensile strength, but greater molding warpage and increased susceptibility to stress cracks. (See how these characteristics vary together in each generic plastic listed in Chap. 4.)
3. Greater material shrinkage, more difficulty in holding tolerances.
4. Less impact strength.
5. Improved heat properties (5).

3.4 MOLECULAR WEIGHT

Molecular weight and the dispersion of different molecular weights in a polymer also have effects on the characteristics of polymers. For example, higher-molecular-weight polymers tend to be tougher and to have increased chemical resistance. Lower-molecular-weight polymers are generally weaker and more brittle (5).

The subject is complex, however; further comments will be of little value in the present volume since information on molecular weight is not generally made available in the published polymer handbooks. For these reasons, we will not cover the effects of molecular weight except for the following remarks:

Ultrahigh-molecular-weight (UHMW) polymers are not viable materials for ultrasonic insertions of fasteners because of problems with melt viscosities.

Ultrahigh-molecular-weight polymers are usually in the olefin family. Torque values required to form threads with some types of self-threading screws in UHMW materials are not usually any greater than the required torque with standard molecular weight polymers.

3.5 MOISTURE CONTENT

Some plastics readily pick up moisture from the atmosphere. Nylon is such a material. The hygroscopic nature of these plastics can cause significant differences in the threading and general characteristics under dry conditions versus when the material has been stabilized at a high ambient humidity.

Atmospheric humidity is constantly changing and the moisture content of such material will tend to change with it. Therefore, it seems generally advisable in the laboratory analysis of the fastening characteristics of relatively high moisture-pickup materials to work with dry materials. For this reason, the resin producers we have worked with on such products ship the material in plastic bags containing a desiccant. The moisture acts as a plasticizer. Expected performance changes with added moisture content are in the increased degree of elasticity and ductility, as well as in dimension and reduction of stiffness. There should be less effect with filled product.

As a crude example of the effect of moisture pickup, we noticed the unusual flexibility of the nylon bristles on two hairbrushes. Examination showed they had been washed, and, in fact, had been soaked for some time. However, within a few days in the dry atmosphere of our home the bristles recovered their usual stiffness.

With increased moisture content (unaccompanied by added filler content) we should expect to find lower torque values and less tendency toward material cracking.

3.6 ELASTICITY AND RESILIENCY IN PLASTICS

3.6.1 Creep

Many qualified people have said that the really big difference between plastics and metals comes in the far greater "creep" that occurs in plastics. In using the word "creep" we refer to the greater capacity of plastics to continually deform under loads over a period of time. This characteristic is increased under higher temperatures.

With plastics, creep varies greatly under different environments, such as temperature, humidity, chemical atmosphere, etc. Therefore when graphically illustrated over a period of time, it is measured under constant conditions and at constant load.

Examination of creep-test figures will show that the creep rate follows a semicurvilinear pattern, similar to that shown in Fig. 3.3. As you can see, the largest deformation occurs in the initial stages with a flattening of the curve in the process of time. However, the deformation under load or stress continues inexorably in all cases over the months and years until the point of rupture. Designers of products manufactured from plastics can use developed information on creep when thinking about the theoretical lifetimes of the various materials they consider.

The creep of plastics can also be illustrated using logarithmic time curves which result in a nearly straight line. We have done a log time graph (Fig. 3.4) on the same material shown in Fig. 3.3 using five-cycle paper to illustrate.

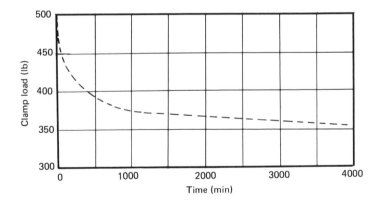

FIG. 3.3   Idealized creep pattern showing 70% retained clamp load over a 64-hr period of a #8-11 Plastite 45 trilobular screw in an engineering grade, acetal-based resin of high strength and good thread-forming ability. Screw engagement length was 0.375 in. The drilled pilot hole at 0.125 in. diameter provided 100% depth of thread in the plastic.

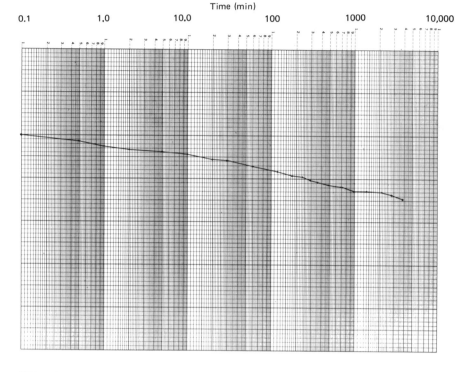

Time (min)

FIG. 3.4   The creep of plastics illustrated in a nearly straight line
using 5-cycle logarithmic graph paper on which the clamp loads have
been recorded over a 64-hr period.

Recorded Clamp Load

Load (lb)	Elapsed time	Load (lb)	Elapsed (hr)
500	0.1 min	396	5
486	0.5	392	6
474	1.0	387	8
468	2.0	384	10
462	5.0	381	12
456	10.0	372+	16
444	20.0	372	24
441	30.0	370	36
429	60.0	364	48
418	2 hr	360	56
408	3	354	64
404	4		

Creep can occur under different kinds of forces such as tension, compression, shear, or flexure; the first three factors should be of especial concern in fastening plastics with mechanical fasteners.

3.6.2 Cold Flow and Resiliency

"A plastic is said to 'cold-flow' when it does not return to its original dimensions after the application of stress. (6) The term 'creep' is sometimes used" interchangeably with "cold flow" (6). Although plastics are relatively resilient and elastic in comparison with metals, they vary considerably in their resiliency, depending on material and filler content. The ability of plastics to return to their original form varies not only with their innate structure and filler content, but also with the degree of stress, elapsed time they have been under stress, and the environmental conditions. Practically speaking, there is usually an increment of nonrecoverable deformation which still exists after immediate partial recovery and further long-term recovery.

The modulus of elasticity in a plastic is "a measure of the ratio of the stress to the strain in a material, within the elastic limits of the material" (6).

3.6.3 Material Relaxation

Under a constant strain, the induced stress will decay with time, and this decay has the effect of reducing the holding power of a mechanically fastened joint. This is referred to as stress relaxation.

Clamp load in a joint is the force holding the joint together. With unmodified polyolefin plastics such as polyethylene or polypropylene a large amount of the initial force holding the joint together is going to be lost very quickly due to stress relaxation, and more will be lost over a period of time. Incidentaly, zinc diecastings have a similar tendency. Some people may recall that in the older-style automobiles, screws used with zinc die-cast mirror-attachment arms used to require occasional retightening.

The effect is far greater with threaded fasteners used in plastics. When clamp load is reduced in a vibrational environment a threaded fastener may tend to rotate and loosen. Examples of the reduction in clamp load from material relaxation with Delrin and polypropylene are shown in Table 3.1.

What to do to rectify loosened joints is explained in Chap. 6, Sec. 6.4.3.

3.7 CRAZING

Crazing is a normal phenomenon with all thermoplastics. The degree of crazing differs with the various materials, as individually covered in the next chapter. Crazing can result from a variety of causes in the manufacturing process or in the design of the piece associated with relief of internal or external stresses. Polymers with higher molecular

TABLE 3.1 Clamp-Load Retention in Plastics

Screw type[a]	Material	Initial clamp load	Test duration - load retention				Pullout (lb)	Mode of failure
			30 sec	10 min	1 hr	12 hr		
Plastite, 45°	Poly-Pro	275	252	210	198	173	620	Hole stripped
	Delrin	500	486	456	429	381	1125	Screw broke

[a]Thread diameter: 0.164−0.160 in.; material thickness: 3/8 in.; hole size: 0.125 in.

weights, lower crystallinity, and less molecular side branching exhibit greater resistance to crazing. Application of excess seating torque for screws used with plastics particularly subject to crazing should be avoided. "There is a minimum or critical strain which is required for crazing to develop in a polymer. If the strain is below this value than crazing will not occur" (7). Counterboring of pilot holes tends to disperse the stress created in driving screws and may alleviate the creation of crazing—*counterboring*, not countersinking.

3.8 ANNEALING AND THERMAL TREATMENTS

The creep rate and stress relaxation of polymers is greatly affected by postproduction treatments such as annealing or quenching at temperatures below the glass transition temperatures (in the case of amorphous or partially amorphous plastics) or the melting point. The glass transition temperature is associated with amorphous or partially amorphous plastics, and it is the temperature below which the polymer behaves like a solid.

Annealing reduces the rate of creep and the rate of stress relaxation. With "crystalline polymers the cause of the decreased creep must be associated . . . with changes in the crystalline state while similar effects in amorphous polymers are largely due to changes in free volume or density" (4).

3.9 CONCLUDING THOUGHTS

We have just barely mentioned this, but the mechanical characteristics and other properties of plastic products are very much affected by manufacturing methods and postproduction treatments. For example, "the extent of crystallinity depends on the rate of cooling through the freezing range," and that is affected by "mold configuration, tool material and wall thickness." Under different manufacturing conditions "chemical configurations which yield identical tensile strengths may not produce identical creep rates, fatigue lives, etc. The properties of the finished product from identical starting materials will therefore vary greatly when produced by different processes, or if manufacturing conditions change" (7). Moral: Don't switch suppliers and molders, once you've started production—if you can avoid it! Keep in touch with the molder if you're the designer! Two laboratory studies in Chap. 7 explain why (see Sec. 7.4 and 7.9).

REFERENCES

1. R. L. Reynolds, *Assembly Engineering* 14:28 (September 1971).

2. E. H. Andrews, Structure-property relationships in a polymer, International Journal of Polymeric Materials 2:337-359 (1973).

3. R. L. E. Brown, *Design and Manufacture of Plastic Parts*, Wiley, New York, 1980, p. 9.

4. L. E. Nielson, *Mechanical Properties of Polymers and Composites*, vol. 1, Marcel Dekker, New York, 1974, pp. 26, 94-95.

5. J. Frados (Ed.), *Plastics Engineering Handbook of the Society of the Plastics Industry, Inc.*, 4th Ed., Van Nostrand Reinhold, New York, 1976, pp. 109-110, 44.

6. R. D. Beck, *Plastic Product Design*, Van Nostrand Reinhold, New York, 1980, pp. 398, 402.

7. E. Miller, *Plastics Products Design Handbook*, part A, Marcel Dekker, New York, 1981, pp. 11-12, 24-25.

BIBLIOGRAPHY

Agranoff, J., Ed., *Modern Plastics Encyclopedia*, Vol. 57 No. 10A, Vol. 58, No. 10A, McGraw-Hill, New York, 1981 and 1982.

Miller, E., Ed., *Plastics Products Design Handbook*, part A, Marcel Dekker, New York, 1981.

Reynolds, R. L., Solutions to Plastic Fastening Problems, *Assembly Engineering* 14:26-30 (September 1971).

Rodriguez, F., *Principles of Polymer Systems*, McGraw-Hill, New York, pp. 26-47.

4
GENERIC NAME PLASTICS

This chapter provides general information on the fastening characteristics of a majority of the polymers with which threaded metal fasteners
can be used. It is thus intended to be only of supplementary assistance to the part designer, when making the complex decisions on material, and to others involved in the decision-making process regarding
fastener selection. Very obviously, the information is inadequate for
complete material selection.*

A reminder should be made at this point of the remarks in Chap. 3
and other portions of this volume, in which we cover the effects of
stiffness and elasticity (as a result of reinforcements, moisture, molecular weight, crystallinity, creep, cold flow, and material relaxation) on
the performance of threaded fasteners in plastics.

As we previously remarked, the ability of a plastic to accept metal
threads without cracking, the torque levels which will be received,
and the general strength of the fastening site are profoundly affected
by the degree of stiffness of the plastic. The flex modulus is a measure of stiffness. Added moisture content increases the elasticity and
reduces stiffness.

High molecular weight increases toughness, resistance to creep, and
stress cracking. Ultrahigh-molecular-weight plastics are unsuitable
for ultrasonic insertion. Higher crystallinity tends to provide increased
resistance to stress cracking. Crystallinity is never complete; there

---

*The majority of the mechanical and other property values shown under
the various materials covered in this chapter were obtained from *Modern
Plastics Encyclopedia*, vol. 58, No. 10A, McGraw-Hill, New York, 1981–
1982; and are reproduced by permission. In some cases, the data were
obtained directly from the resin producer and are so marked.

are always some amorphous areas. Density is one of the measures of crystallinity. In this chapter we have listed the crystalline and the amorphous thermoplastic polymers separately.

A list of tradename plastics which are cross-referenced in Chap. 5 appear at the end of this chapter.

The American Society of Testing and Materials (ASTM) Test Methods* used for the properties shown in this chapter are:

    Flexural modulus, ASTM D790
    Tensile modulus, ASTM D638
    Tensile strength at break, ASTM D638
    Mold linear shrinkage, ASTM D955
    Specific gravity, ASTM D792
    Hardness: Rockwell, ASTM D785
               Shore, ASTM D2240
    Water absorption (% wt), ASTM D570

Fastening characteristics described here are the opinion of the authors, resulting from testing in all cases but one by Kenneth J. Gomes and others at the fastener laboratory of Continental/Midland (formerly Continental Screw) or its fastener-licensing facility, Research Engineering and Manufacturing, all now relocated to Park Forest, Illinois. Polyethersulfone was tested for the authors at the fastener laboratories of Elco Industries, Inc., Rockford, Illinois.

THERMOPLASTIC RESINS, CRYSTALLINE

4.1 POLYOLEFINS

The prime difference in fastening characteristics between the various olefins stems from variations in the degree of crystallinity, causing significant differences in tensile, flexural, and impact strengths, as well as resistance to stress cracking (1).

4.1.1   Polyethylene (PE): Low-Density (LDPE), High-Density (HDPE), and Ultrahigh-Molecular-Weight (UHMWPE)

Properties and Characteristics

Polyethylene is a lightweight, low-cost, and easily processed material, chemically resistant and with low water permeability; it has toughness

---

*Some ASTM test methods are described in the *Plastics Engineering Handbook of the Society of Plastics Engineers*, 4th Ed. (J. Frados, Ed.), New York, 1976, pp. 872-888.

and good electrical resistance; low heat resistance, and is available in
a wide range of densities. Low-density products are tough and flexible. The high-density product (HDPE) provides higher strength,
stiffness, creep resistance, and improved heat resistance.

Fastening characteristics
Torque. Low-density material produces low drive and fail torques.
Tensile. Low pull-out can also be expected in unfilled grades.
Installations require sufficient engagement length to compensate.
Cracking. Unfilled grades have a high level of ductility. Crack-
ing seldom occurs when using any self-threading fasteners.
Material relaxation. Material has a high level of initial load relaxa-
tion, resulting in 30—50% loss of assembly clamp load in some
grades.

Although in most cases standard thread-forming screws will work
satisfactorily in nonstructural applications, there may be problems with
low pullout in the low-density material. High-density (HDPE) and ultrahigh-molecular-weight (UHMWPE) polyethylene materials are amenable
to the use of special design thread-rolling screws for plastics where,
we feel, they will give superior results. On glass filled product we
recommend testing.

Use of high-density material also produces good results with push-in
thread-forming fasteners especially designed for plastics, and ultrasonically installed, internally threaded metal inserts; however, in
selecting any kind of insert for regular or for ultrasonic installation
it may be wise to consider the relatively low holding power of unreinforced polyolefins.

Ultrasonic insertion is not recommended for ultrahigh-molecular-
weight applications, as the melt viscosity is too high and the UHMWPE
material does not flow easily enough.

Applications. Chemical tanks, toys, cans, boxes, machinery parts,
prosthetic devices, medical instrumentation, etc.
Tradenames. 1900 UHMW, Chemplex, Tenite (PE), Tivar, etc.

Representative suppliers	High density	High molecular weight	Ultrahigh molecular weight
Allied Chemical	x	x	
American Hoechst	x	x	x
American Polymers	x		
Arco	x		
Chemplex	x		
Dow	x		

Representative suppliers	High density	High molecular weight	Ultrahigh molecular weight
Du Pont	x		
Georgia-Pacific	x		
Gulf	x	x	
Hercules			x
Phillips	x	x	
Poly-Hi/Menasha			x
Union Carbide		x	
Others			

Properties	High-density unfilled	High-density 30% glass filled	Ultrahigh molecular weight	Low-density unfilled
Flexural modulus, 10^3 psi @ 73°F	100-260	130-140	up to 800	8-60
Mold linear shrinkage, in./in.	0.015-0.050	0.002-0.006	0.040	0.015-0.050
Tensile strength at break, psi	3100-5500	9000	5600	600-2300
Tensile modulus, 10^3 psi	60-180	—	—	14-38
Specific gravity	0.941-0.965	1.28	0.94	0.910-0.925
Hardness:				
Rockwell	—	R 75	R 50	—
Shore	D 60-70	—	—	D 40-51
Water absorption, % (1/8 in. thick specimen), 24 hr	<0.01	0.02	<0.01	<0.01

Source: Ref. 7.

4.1.2   Polypropylene (PP): Homopolymer or
Copolymer

Properties and Characteristics

Polypropylene generally has good detergent, grease, and chemical resistance; it is harder and more scratch resistant than polyethylene; tough and resistant to stress; heat resistant to 250°F (121°C); electroplatable; and water-vapor resistant.

Copolymer has best impact resistance; mechanical properties of the unmodified product are adversely affected by prolonged heat or solar radiation, but can be improved through the use of additives. Cost is the lowest among the high-volume thermoplastics.

Fastening characteristics. Generally similar to polyethylene.
Torque. Low drive and fail torque values generally achieved.
Tensile. Unfilled grades have low pull-out. Installations require
sufficient engagement to compensate.
Cracking. Good resistance to stress cracks. Cracking seldom occurs with use of self-threading fasteners.
Material relaxation. Initial load relaxation is quite high and 30—50%
loss of assembly clamp load can be expected.

Provided moderation is used on filler content, both homopolymers and copolymers display good thread-forming characteristics with standard thread-forming screws, although we belive that special-design thread-rolling screws for plastics will give superior results. Polypropylene is recommended for use with push-in thread-forming fasteners especially designed for plastics and/or ultrasonic installation of internally threaded metal inserts.

In considering design of ultrasonic inserts or regular inserts, consideration should be given to the relatively low holding power of unfilled polyolefins.

Applications. Coffee makers, vaporizers, washing machines, pumps,
automotive kick panels, fan shrouds, bucket seat backs, battery
covers, auto air-conditioning and heater ducts, air-cleaner housings, distributor caps, etc.
Tradenames. Moplen, Profax, Profil, Rexene (PP), Tenite
(PP), etc.
Representative suppliers. American Polymers, Arco, Exxon, Gulf,
Hercules, Phillips, Rexene, USS Novamont, etc.

Properties	Homopolymer	Copolymer	Talc filled	Glass filled
Flexural modulus, 10^3 psi @ 73°F	170-250	130-200	450-625	950-1000

Properties	Homopolymer	Copolymer	Talc filled	Glass filled
Mold linear shrinkage, in./in.	0.010-0.025	0.020-0.025	0.008-0.015	0.003-0.005
Tensile strength at break, psi	4500-6000	4000-5500	4300-5000	8500-15,000
Tensile modulus, 10^3 psi	165-225	100-170	450-575	1100-1500
Specific gravity	0.900-0.910	0.890-0.905	1.23-1.27	1.22-1.23
Hardness: Rockwell	R80-102	R50-96	R94-R110	R102-111
Water absorption, % (1/8 in. thick specimen): 24 hr saturation	0.01-0.03 —	0.3 —	0.01-0.03 —	0.05-0.06 0.09-0.10

Source: Ref. 7.

4.1.3 Polyallomer

Properties and Characteristics

This is an alloy of polyethylene and polypropylene with unusually high toughness and impact resistance. Improved abrasion resistance over PE or PP. Heat resistant to about 210°F (99°C).

Fastening characteristics. Being an alloy of PE and PP, many of the fastening characteristics are similar.
   Torque. Low drive and fail torque values can be expected.
   Tensile. Pull-out in unfilled grades is rather low. Requires sufficient engagement lengths to compensate.
   Cracking. Improved stress-cracking resistance over PE.
   Material relaxation. Loss of initial, assembly clamp load can be expected in about the same degree as PE and PP.

Amenable to standard thread-forming screws, although we believe that special-design thread-rolling screws for plastic will give superior performance. Recommended with push-in thread-forming fasteners especially designed for plastics and/or ultrasonic installations of internally threaded metal inserts.

In considering the design of inserts for ultrasonic or regular use, consideration should be given to the low holding power of unreinforced polyolefins.

Applications. Notebook covers, luggage, typewriter cases, etc.
Tradename. Tenite (PA)
Representative supplier. Eastman

Properties	PA
Flexural modulus, 10^3 psi @ 73°F	70-110
Mold linear shrinkage, in./in.	0.010-0.020
Tensile strength at break, psi	3000-3800
Tensile modulus, 10^3 psi	—
Specific gravity	0.896-0.899
Hardness: Rockwell	R 50-85
Water absorption, % (18 in. thick specimen), 24 hr	<0.01

Source: Ref. 7.

4.2 HIGH-PERFORMANCE, THERMOPLASTIC CRYSTALLINE POLYMERS

These materials are designed mainly for special and structurally strong applications where the characteristics of polyolefin thermoplastics are inadequate. These materials are usually harder, stronger, and more rigid than the olefins. They are often supplied with glass, mineral, or glass-mineral reinforcements. Structural foams are produced from some of these materials.

4.2.1 Acetal: Homopolymer or Copolymer

Properties and Characteristics

High dimensional stability, heat resistance, strength and toughness; abrasive resistant; ductile; although moderately hygroscopic, the mechanical characteristics are not affected greatly.

Fastening characteristics
Torque. Acetal has low drive torque and high fail torque.
Tensile. The holding power with thread-rolling screws is high,
    similar to that of metal application.
Cracking. Tendency toward stress cracking is low.
Material relaxation. Initial load relaxation tends to be very low.

Acetal (either as a homopolymer or a copolymer) is used very successfully with special-design thread-rolling screws and push-in thread-
forming fasteners for plastics and exhibits good characteristics with
ultrasonic installation of internally threaded metal inserts. Molded-in
inserts are not recommended in unmodified product (2). Section 10.2
contains detailed information on this subject.

Applications. Bearings, bushings, water pump impellers, water closet
    components, automotive instrument cluster, office machine housings,
    lock parts, etc.
Tradenames. Celcon (copolymer), Delrin (homopolymer)
Representative suppliers. Celanese, Du Pont

	Homopolymer		Copolymer	
Properties	Unfilled	20% Glass filled	Unfilled	25% Glass filled
Flexural modulus, 10^3 psi @ 73°F	380-430	730	375	1100
Mold linear shrinkage, in./in.	0.020-0.025	0.009-0.012	0.020 (Av.)	0.004 flow 0.018 trans.
Tensile strength at break, psi	—	8500	—	18500
Tensile modulus, 10^3 psi	520	1000	410	1250
Specific gravity	1.42	1.41	1.56	1.61
Hardness: Rockwell	M94	M98	M78	M79
Water absorption, % (1/8 in. thick specimen), 24 hr	0.25-0.40	0.25	0.22	0.29

Source: Ref. 7.

CELLULOSICS

The cellulosics generally have clarity, provide electrical insulation, and are scuff resistant. Cellulosics must be chemically modified to be classed as a thermoplastic.

4.2.2   Cellulose Acetate Butyrate (CAB)

Properties and Characteristics

Very tough, hard surface; improved impact resistance over CP; easy to clean; weather-resistant; best dimensional, moisture, and chemical resistance of all cellulosics; compatible with some plasticizers; color stable.

Fastening characteristics
   Torque and tensile. Cellulosics tend to have relatively high torque and tensile characteristics.
   Cracking. When used with thread-rolling screws, usually requires larger than normal pilot holes to avoid stress cracking.
   Material relaxation. Very little loss of initial clamp load is expected.

Amenable to special-design thread-rolling screws and push-in thread-forming fasteners for plastics. However, the fastener manufacturer's product engineering laboratory should be consulted. Thread-cutting screws are often recommended for use in cellulosics. CAB is an excellent material for internally threaded metal inserts, ultrasonically installed.

   Applications. Outdoor lighted signs, tail light lenses, steering wheels, storm windows, eyeglasses, handles, etc.
   Tradenames. Celanese CAB, Tenite CAB
   Representative suppliers. American Polymers, Eastman, Mobay, etc.

Properties	CAB molding compound
Flexural modulus, 10^3 psi @ 73°F	90-300
Mold linear shrinkage, in./in.	0.003-0.009
Tensile strength at break, psi	2600-6900
Tensile modulus, 10^3 psi	50-200
Specific gravity	1.15-1.22

Properties	CAB molding compound
Hardness: Rockwell	R 31-116
Water absorption, % (1/8 in. thick specimen), 24 hr	0.9-2.2

Source: Ref. 7.

4.2.3 Cellulose (Acetate) Proprionate (CAP)

Properties and Characteristics

Hard and quite tough; easy to clean, with good weather resistance; compatible with plasticizers; color stable.

Fastening characteristics
   Torque and tensile. Cellulosics can yield relatively high torque
      and tensile capabilities.
   Cracking. When used with thread-rolling screws, usually requires
      larger than normal pilot holes to avoid stress cracking.
   Material relaxation. Very little assembly load relaxation occurs.

Amenable to special-design thread-rolling screws and push-in thread-forming fasteners for plastics. However, consultation should be made in all cases with the fastener manufacturer's product engineering laboratory. Thread-cutting screws are often recommended for use in cellulosics. CAP is an excellent material for ultrasonic installation of internally threaded metal inserts.

   Applications. Telephone housings, transistor radio housings, steer-
      ing wheels, eyeglass frames, safety goggles and shields, motor
      covers, etc.
   Tradenames. Tenite CP
   Representative suppliers. American Polymers, Eastman, etc.

Properties	CAP molding compound
Flexural modulus, 10^3 psi @ 73°F	120-350
Mold linear shrinkage, in./in.	0.003-0.009
Tensile strength at break, psi	2000-7800

Properties	CAP molding compound
Tensile modulus, 10^3 psi	60-215
Specific gravity	1.17-1.24
Hardness: Rockwell	R10-122
Water absorption, % (1/8 in. thick specimen), 24 hr	1.2-2.8

Source: Ref. 7.

4.2.4 Fluoropolymers: (poly)Tetrafluoroethylene (TFE/PTFE) and Fluoronated Ethylene Propylene (FEP)

Use of mechanical fasteners in all fluoroplastics should be subject to tests by a competent fastener-testing laboratory.

Properties and Characteristics

Excellent resistance to heat, chemicals, weather and impact; high impact resistance; nonburning; very low water absorption; resilient with very low coefficient of friction. FEP is the easiest fluoroplastic to process.

Fastening characteristics
> Torque and tensile. Fastening site must be properly designed and attention paid to screw selection to obtain satisfactory torque and tensile levels.
> Cracking. For most grades, ductility is high and material resists cracking.
> Material relaxation. High cold-flow characteristics must be offset with deep thread-engagement lengths. Periodic retightening of screws is often required, especially with the unmodified product owing to the extreme level of load relaxation. We recommend that consideration be given to the modified grades if threaded fasteners are to be used.

Molded applications accept the use of thread-rolling screws quite readily, but the material is so resilient and lubricious that testing and hole-size recommendation by a competent fastener laboratory are an absolute "must." Only threaded fasteners with the highest available prevailing torque characteristics should be considered for use with any unmodified product. We suggest the special-design, double-lead, trilobular thread-rolling screws for plastics. Use of push-in types of

thread-forming fasteners is not recommended. Ultrasonic installations
are not recommended for TFE or FEP, but there are melt-processible
grades which provide varying results.

Applications. Bearings, bushings, slides, pumps, impellors, valves,
    etc.
Tradenames. Fluon, Halon, Teflon
Representative suppliers. Allied Chemical, Du Pont, ICI Americas

Properties	TFE granular	TFE-25% glass filled	FEP
Flexural modulus, 10^3 psi @ 73°F	80	235	80-95
Mold linear shrinkage, in./in.	0.030-0.060	0.018-0.020	0.030-0.060
Tensile strength at break, psi	2000-5000	2000-2700	2700-3100
Tensile modulus, 10^3 psi	58-80	—	50
Specific gravity	2.14-2.20	2.2-2.3	2.12-2.17
Hardness: Rockwell	—	—	—
Shore	D50-55	D60-70	D60-65
Water absorption, % (1/8 in. thick specimen), 24 hr	<0.01	—	<0.01

Source: Ref. 7.

4.2.5 Nylon (Polyamide)

Properties and Characteristics

Nylon is produced in a number of grades, designated numerically by
chemical structure, for example, Nylon 6, 6/6, 6/12, 11, 12. It lends
itself well to reinforcements, fillers, and blends and is characterized
by excellent wear resistance, toughness, and impact strength; has
good chemical resistance; quite hygroscopic.

Fastening characteristics. Because of the hygroscopic nature of
    Nylon, fastening characteristics may vary considerably with changes
    in humidity.
        Torque and tensile. Under dry conditions, relatively high torque
        and tensile levels can be expected. However, as moisture con-
        tent rises, test values will be lower.

Cracking. At reasonable reinforcement levels, the material normally has good resistance to stress cracking.
Material relaxation. Filled material is not normally subject to detrimental load relaxation. Unfilled grades maintain loads well.

Nylon is an excellent material for special-design thread-rolling screws and push-in thread-forming fasteners for plastics. Good characteristics are displayed with threaded metal inserts, ultrasonically installed. As with other hygroscopic materials, insertion should be done right after molding or the material should be stored in sealed containers with an appropriate dessicant. Otherwise, voids and fissures may result.

Applications. Gears, pumps, valves, bushings, tool housings, windshield-wiper parts, appliance housings, impellers, chemical pumps, valve covers, electric plugs, etc.
Tradenames. Capron, Durathane, Fosta, Maranyl, Minlon, Nylafil, Nypel, Ultramide, Vydyne, Wellamid, Zytel, etc.
Representative suppliers. Allied Chemical, American Hoechst, Celanese, Du Pont, Fiberfil, Chemical, Monsanto, etc.

	Nylon 6			
Properties	Molding compound	30-35% Glass filled	40% Mineral and glass fiber filled	High-impact copolymer
Flexural modulus, 10^3 psi @ 73°F	390[a] 140[b]	1500[a] 800[b]	1300[a] —	110-320[a] —
Mold linear shrinkage, in./in.	0.005-0.015	0.004	0.004-0.006	0.008-0.018
Tensile strength at break, (psi)	18,000[a] 10,000[b]	25,000[a] 13,000[b]	16,000[a]	7500-11,000
Tensile modulus (10^3 psi)	380[a]	1450[a]	—	—
Specific gravity	1.12-1.14	1.35-1.42	1.46-1.50	1.08-1.17
Hardness: Rockwell	R119[a] —	M101[a] M78[b]	R118[a]	R81-110[a] —

Properties	Molding compound	30-35% Glass filled	40% Mineral and glass fiber filled	High-impact copolymer
Water absorption, % (1/8 in. thick specimen):				
24 hr	1.3-1.9	1.2	0.9	1.3-1.5
Saturation	8.5-10.0	6.5-7.0	6.0	8.5

[a]Dry, approximately 0.2% moisture.
[b]Conditioned to equilibrium at approximately 50% relative humidity.
Source: Ref. 7.

	Nylon 6/6			
Properties	Molding compound	Impact modified	33% Glass filled	Mineral filled
Flexural modulus, 10^3 psi @ 73°F	420[a] 185[b]	275[a] 150[b]	1300[a]	900[a] 400[b]
Mold linear shrinkage, in./in.	0.008-0.015	0.013-0.018	0.005	0.012-0.022
Tensile strength at break, psi	12,000[a] 11,000[b]	7000-8000	28,000[a] 22,000[b]	14,000[a] 11,000[b]
Tensile modulus, 10^3 psi	—	—	—	900[a] 500[b]
Specific gravity	1.13-1.15	1.08-1.10	1.38	1.39-1.47
Hardness: Rockwell	R120[a] M83[b]	R114-115[a] —	M100[a] —	R106-119[a] —

Properties	Molding compound	Impact modified	33% Glass filled	Mineral filled
Water ab- sorption, % (1/8 in. thick speci- men):				
24 hr	1.0-1.3	—	1.0	0.6-0.9
saturation	8.5	—	6.5	6.0-6.5

[a]Dry, approximately 0.2% moisture.
[b]Conditioned to equilibrium at approximately 50% relative humidity.
Source: Ref. 7.

4.2.6 Thermoplastic Polyester: Polybutylene Terephthalate (PBT) and Polyethylene Terephthalate (PET)

Properties and Characteristics

Thermoplastic polyester is a replacement for many thermoset applications, providing improved impact resistance; low moisture absorption; chemical resistance; excellent electrical properties; and is readily machined (minimal warpage). Also available as structural foam.

Thermoplastic polyesters are characterized by an ability to be molded into very thin sections. These materials are currently replacing thermoset plastics such as phenolics in many electrical applications because of their high dielectric strength and tracking resistance. Some grades can be manufactured with a small amount of blowing agent to create a small density reduction for even greater stiffness. Stiffness and rigidity are the most important characteristics of this resin.

Fastening characteristics
Torque. In both PET and PBT, test results show moderate drive torques accompanied by fail torques giving a fail-drive ratio from approximately 3:1 to a maximum of about 5:1 with a major fastener company's special design roll-thread screws for plastics.
Tensile. With these same screws, pullout tests for the #8 size screws with 7/16—1/2 in. engagement length in PBT showed resistance in excess of 1000 lbs and proportionately higher values for PET very much in accordance with expected results, allowing for the material differences.
Cracking. Notch-sensitive. Fastening site (boss) should take this tendency into consideration.
Material relaxation. Has very low tendency toward load relaxation.

Amenable to special-design thread-rolling screws·and push-in thread-

forming fasteners for plastics. On certain applications, standard thread-
cutting screws may be recommended, but they should never be used
with structural foam versions. Has good characteristics for ultrasonic
installation of internally threaded metal inserts.

>   *Applications.* Automotive ignition and body parts, electrical com-
>   ponents, high-voltage TV components, replaces brass in some
>   marine applications, appliance housings, pumps and impellers,
>   gasoline-powered tool housings, etc.
>   *Tradenames.* Celanex, Gafite, Gaftuf, Petlon, Petra, Rynite, Valox,
>   Tenite polyterephthalate
>   *Representative suppliers.* Allied, Celanese, DuPont, GAF, General
>   Electric, Mobay

Properties	PBT		PET	
	Unfilled	30% Glass filled	Unfilled	30% Glass filled
Flexural modulus, 10^3 psi @ 73°F	330-400	1100-1200	350-450	1300
Mold linear shrinkage, in./in.	0.015-0.020	0.002-0.008	0.020-0.025	0.002-0.009
Tensile strength at break, psi	8200	17,000-19,000	8500-10,000	23,000
Tensile modulus, 10^3 psi	280	1300	400-600	1440
Specific gravity	1.31-1.38	1.52	1.34-1.39	1.57
Hardness: Rockwell	M68-78	M90	M94-101	M100
Water ab- sorption, % (1/8 in. thick speci- men), 24 hr	0.08-0.09	0.06-0.08	0.1-0.2	0.05

Source: Ref. 7.

4.2.7 Polyphenylene Sulfide (PPS)

Although classed as a thermoplastic, depending on cross-linkage in molding cycle, linkage can reach thermoset point.

Properties and Characteristics

Good thermal stability and excellent chemical resistance; good mechanical properties while classed as a thermoplastic; good flame resistance; heat deflection temperature approximately 500°F (260°C); good electrical properties.

Fastening characteristics

Torque. PPS has relatively high drive and fail torque levels.

Tensile. Tensile (pull-out) level is usually quite high.

Cracking. Subject to stress cracks and notch sensitivity; counter-bore pilot holes.

Material relaxation. Initial load relaxation can be expected to be quite low.

When classed as a thermoplastic, it can be used for special-design thread-rolling screws for plastics. Not recommended for push-in thread-forming fasteners. Good suitability for ultrasonic installation of internally threaded metal inserts.

Applications. Electrical and electronic components, chemical pumps, appliance housings, under-the-hood automotive components, hair dryers, cooking appliances, etc.

Trademark. Ryton

Supplier. Phillips

Properties	Injection molding[a]	40% Glass filled	Mineral and glass filled
Flexural modulus, 10^3 psi @ 73°F	550	1700	1800
Mold linear shrinkage, in./in.	0.006-0.008	0.002-0.004	—
Tensile strength at break, psi	9500	19,500	13,000
Tensile modulus, 10^3 psi	480	1100	—
Specific gravity	1.3	1.6	1.8
Hardness: Rockwell	R123	R123	R121

Properties	Injection molding[a]	40% Glass filled	Mineral and glass filled
Water absorption, % (1/8 in. thick specimen), 24 hr	<0.02	0.05	0.03

[a]Although the properties are listed for unreinforced material, it is our understanding that PPS is available only in reinforced versions (3).
Source: Ref. 7.

THERMOPLASTIC RESINS, AMORPHOUS

4.3 VINYLS

4.3.1 Polyvinyl Chloride (PVC)

Properties and Characteristics

Can be rigid or flexible, depending on the addition of plasticizers; however, threaded fasteners are not generally used with plasticized product. Resistant to weather, moisture, and chemicals, but susceptible to heat degradation; heat deflection temperature (HDT) 140−170°F (60−77°C) in rigid form; low cost.

Fastening characteristics
   Torque. Plasticized grades yield low torque values. Moderate
      torques can be expected in unplasticized materials.
   Tensile. Pullout level is conditional with the same circumstances.
   Cracking. Plasticized grades, being ductile, resist cracking.
      Some unplasticized grades may require care with fastening-site
      design to avoid cracking.
   Material relaxation. Plasticized and low-strength grades react
      like the olefins, in that 30−50% assembly clamp-load loss can be
      expected. Stiff rigid grades provide good assembly load reten-
      tion.

Unplasticized, rigid PVC is a very satisfactory material for use with either standard thread-forming or special-design thread-rolling screws for plastics. Also highly suitable for push-in thread-forming fasteners especially designed for plastics, and gives superior results with ultrasonic installation of internally threaded metal inserts. Plasticized PVC is not recommended for use with ultrasonic assembly owing to the fact that high amounts of plasticizer and stabilizer result in thermal degradation and discoloration.

Applications. (unplasticized). Pumps, impellers, window channels, molding trim, etc.

Tradenames. Geon, Marvinol, PVC-CON, Pliovic, Rocor, Rucoblend, etc.

Representative suppliers. (unplasticized). Conoco, Diamond Shamrock, Ethyl, Goodrich, Occidental, Stauffer, Uniroyal, etc.

Properties	Std. rigid mold. compd.	PVC mld. cmpd. 15% glass filled	Chlorinated PVC
Flexural modulus, 10^3 psi @ 73°F	300-500	750	300-450
Mold linear shrinkage, in./in.	0.001-0.005	0.001	0.003-0.007
Tensile strength at break, psi	6000-7500	9500	7500-9000
Tensile modulus, 10^3 psi	350-600	870	360-475
Specific gravity	1.30-1.58	1.54	1.49-1.58
Hardness: Rockwell	—	R118	R117-122
Shore	D65-85	—	—
Water absorption, % (1/8 in. thick specimen), 24 hr	0.04-0.4	0.01	0.02-0.15

Source: Ref. 7.

4.4 HIGH-PERFORMANCE THERMOPLASTIC AMORPHOUS POLYMERS

These materials are used for more structural applications where the vinyls do not suffice. The materials are usually harder, stronger, more rigid or they have other desirable characteristics. They are often supplied with glass, mineral, or glass/mineral reinforcements. Structural foams are produced from some of these materials.

4.4.1 Acrylics

Properties and Characteristics

Good wear resistance; fair chemical resistance; excellent electrical properties; hard surface; often modified with PVC for high impact resistance; usually transparent.

Fastening characteristics
Torque. High drive and fail torque values can be expected.
Tensile. Material has relatively high clamp load and tensile, pull-
out capabilities.
Cracking. Acrylics are notch sensitive and subject to cracking.
Fastening site (boss) must be designed to withstand the tendency
toward stress cracking.
Material relaxation. Initial load relaxation tends to be low.

Amenable to special-design thread-rolling screws and push-in thread-
forming fasteners for plastics. Unmodified acrylic, multipolymer xt
polymer, as well as PVC and impact-modified acrylics have good ratings
for use with internally threaded metal inserts, ultrasonically installed.

Applications. Automotive tail light lenses and reflectors, aircraft
windows, skylights, outdoor lighted signs, knobs, etc.
Tradenames. Acrylite, Lucite, Plexiglas, etc.
Representative suppliers. Du Pont, Rohm and Haas, etc.

	Molding compound	
Properties	Impact modified	Heat resistant
Flexural modulus, 10^3 psi @ 73°F	200-380	460-500
Mold linear shrinkage, in./in.	0.004-0.008	0.004-0.007
Tensile strength at break, psi	5000-9000	10,000
Tensile modulus, 10^3 psi	200-400	350-460
Specific gravity	1.11-1.18	1.16-1.19
Hardness: Rockwell	R105-120	M95-105
Water absorption, % (1/8 in. thick speci- men), 24 hr	0.2-0.8	0.2-0.3

Source: Ref. 7.

4.4.2   Polycarbonate (PC)

Properties and Characteristics

Replaces zinc die castings; excellent outdoor and dimensional stability;
rigid; superior toughness; heat resistant 250-300°F (121-149°C); low

water absorption; good optical clarity; also available as structural
foam.

Fastening characteristics
   Torque and tensile. With proper lengths of engagement, torque
   and tensile values can approach that of metals. An example is
   shown in the laboratory study in Chap. 8, Sec. 8.6.
   Cracking. With the nonfoam product, fastener-site design is
   critical because of possible internal stresses. With thread-rolling
   screws, stress cracking can be avoided in the unreinforced material if molding stress can be kept no higher than approximately
   1200 psi. The use of thread-rolling screws will add a minimum
   stress of 300 psi at the fastener site. Structural foam polycarbonates are not as critical, as mold stress creation is not a
   factor.
   Material relaxation. Tendency is low, especially when reinforced.

   In many cases, polycarbonate designs are very amenable to the use
of special-design thread-rolling screws or possibly, trilobular design
threaded push-in fasteners for plastics, but a competent fastener laboratory should be consulted. In some cases, use of thread-cutting
screws may be recommended, but they should never be used with structural foam.
   Ultrasonically installed, internally threaded metal inserts can give
good results with proper procedure and are recommended in place of
molded-in metal inserts, which in larger sizes may require preheating
(4).*

Applications. Can be used as a direct replacement for some zinc die
   castings; PC is used for street-lamp housings, lenses, office machine housings, pump impellers, appliance housings, shatter-re-
   sistant lenses and shields, storm and school windows, automotive
   instrument panels and bumpers, etc.
Tradenames. Lexan, Merlon, Polycarbafil
Representative suppliers. Fiberfil, General Electric, Mobay

Properties	Unfilled	10% Glass filled	30% Glass filled	PC-ABS alloy
Flexural mod- ulus, 10^3 psi @ 73°F	340	500	1100	300-400

---

*Note. With ultrasonic insertions, attention should be given to the
sensitivity of polycarbonates for pick-up of moisture at processing
temperatures.

Properties	Unfilled	10% Glass filled	30% Glass filled	PC-ABS alloy
Mold linear shrinkage, in./in.	0.005-0.007	0.002-0.005	0.001-0.002	0.005-0.009
Tensile strength at break, psi	9500	9500	19,000	7000-7300
Tensile modulus, 10^3 psi	345	500	1250	370-380
Specific gravity	1.2	1.27-1.28	1.4	1.12-1.20
Hardness: Rockwell	M70	M75	M92	R117
Water absorption, % (1/8 in. thick specimen), 24 hr	0.15	0.15	0.14	0.21-0.24

Source: Ref. 7.

4.4.3 Polyetherimide

Properties and Characteristics

Transparency; high heat resistance; strengths beyond most engineering resins and at higher usable temperatures, unmodified product has heat deflection temperature of 392°F (200°C); resistance to wide range of chemicals and solvents; good electrical properties; low water absorption; unusually ductile for such a high-strength material and retains considerable ductility even with 10% glass reinforcement; resistant to ultraviolet and radiation exposure; machineable.

> *Fastening characteristics*
> Torque. Low to moderate drive torque and quite high fail torque should be expected.
> Tensile. Holding power of special-design thread-rolling screws for plastic in this material is excellent.
> Cracking. Tends to be notch sensitive, so that the less stress created by the screw design, the better. Counterbore of pilot holes is desirable.
> Material relaxation. Low initial load relaxation to be expected.

With the elasticity of this material, special-design thread-rolling screws for plastic form good, strong, uniform threads at low-to-moderate drive torques. Because of the high failure torque which has been demonstrated under laboratory test conditions these screws should give excellent fail-drive ratios and differentials. The tests show excellent strength retention after repeated removal and reinsertion or under vibration. Thus, they can be recommended for use in high-speed assembly and for automotive use.

Amenable to molded-in, threaded inserts at high molding temperatures where the design is kept simple and knurls are minimal. As a starting point for the users own independent testing some ratios of wall thickness are shown below. Users should read the resin manufacturer's publications, statements, and any cautions about product usage. (See Ref. 5.)

Insert material	Ratio of wall thickness to OD
Steel	1.0
Brass	0.9
Aluminum	0.8

Ultrasonic insertion of threaded inserts is highly recommended, as less residual stress is created under the correct procedure with a uniform melt and minimal thermal shrinkage. Push-in threaded fasteners for plastics should be inserted by ultrasonic methods only.

Applications. Aerospace components, automotive parts
Tradename. Ultem
Supplier. General Electric Company

Properties[a]	Unfilled	10% Glass filled	20% Glass filled	30% Glass filled
Flexural modulus, 10^3 psi @ 73°F	480	650	900	1200
Mold linear shrinkage, in./in.	0.5-0.7	0.4	0.2-0.3	0.2
Tensile strength at break, psi[a]	15,200 (@ yield)	16,600 (@ failure)	20,100 (@ failure)	24,500 (@ failure)

Properties[a]	Unfilled	10% Glass filled	20% Glass filled	30% Glass filled
Tensile modulus, 10^3 psi	430	650	1000	1300
Specific gravity	1.27	1.34	1.42	1.51
Hardness: Rockwell	M109	M114	M118	M125
Water absorption, % (1/8 in. thick specimen):				
24 hr	0.25	0.28	0.26	0.18
saturation	1.25	1.0	1.0	0.9

[a]Data from General Electric Co.

4.4.4   Polyimide (PI)

Properties and Characteristics

Excellent physical and outstanding electrical properties along with exceptionally high heat resistance; strong; chemical resistance to most organic solvents, dilute acids, lubricants and fuels; somewhat subject to weathering, and loss of some tensile strength may result from prolonged exposure; relatively high cost.

Fastening characteristics

Torque. Drive-torque values are moderate; fail-torque values are quite high, so that drive-to-fail ratios and differentials between drive and fail appear to be highly favorable to the use of special-design thread-rolling screws for plastic.

Tensile. Observed tests displayed high initial clamp loads and substantial pull-out values.

Cracking. The manufacturer reports the material to be similar to acrylics in relation to stress cracking. Since acrylics tend to be somewhat notch-sensitive, design of the fastening site should take this tendency into consideration. Pilot holes should be counterbored to help relieve potential stress.

Material relaxation. Quite high resistance to creep.

Amenable to special-design thread-rolling screws. Push-in thread-forming fasteners for plastics might possibly be usable; however, these are usually recommended for nonstructural applications.

Observed tests show that the material produces an extremely stringy chip when drilled; therefore, a competent fastener laboratory should be consulted before using thread-cutting fasteners. Neither ultrasonic installation nor induction heat installation methods for inserts nor push-in fasteners should be attempted, since the material has no melting point.

Applications. Component parts for aircraft, space hardware; high-reliability printed circuit boards; hot, corrosive, and nuclear applications. With graphite filler and, in some grades also with Teflon, it is used for bearings and other applications subject to wear.
Tradename. Vespel
Supplier. Du Pont

Properties	Unfilled	40% Graphite filled
Flexural modulus, 10^3 psi @ 73°F	450-500	M A
Mold linear shrinkage, in./in.	—	I N
Tensile strength at break, psi	10,500-17,100	U S E
Tensile modulus, 10^3 psi	300	F O R
Specific gravity	1.36-1.43	B
Hardness: Rockwell	E 52-99	E
Shore		A R
Water absorption, % (1/8 in. thick specimen):		I N
24 hr	0.24	G
saturation	1.2	S

Source: Ref. 7.

4.4.5 Modified (poly)Phenylene Oxide (PPO)

Properties and Characteristics

Noryl is a PPO which has been modified for better processing. It is very tough and displays superior dimensional stability with high chemical resistance; heat-resistant to 220°F (104°C); low temperature impact resistance; low water absorption; excellent electrical properties. Also available as structural foam.

Fastening characteristics
Torque and tensile. With proper lengths of engagement, torque
and tensile values can approach that of metals.
Cracking. With the nonfoam product, fastener-site design is
critical because of possible internal stresses. Structural foam
modified phenylene oxide is not as critical, as mold stress creation is not normally a factor.
Material relaxation. Tendency is low, especially when reinforced.

PPO is amenable to the use of special-design thread-rolling screws
and push-in thread-forming fasteners for plastics. Because of the
critical nature of site design, a competent fastener laboratory or the
resin producer should be consulted. In some cases, owing to design
features, thread-cutting screws may be recommended, but they should
never be used with structural foam. PPO has excellent suitability for
use of internally threaded metal inserts, ultrasonically installed.

Applications. Electrical parts, panels and business machine housings; hot water pumps and impellors; utensils and containers;
automotive panels, grilles, dashboards; television components; etc.
In either rigid or foamed versions, Noryl is very popular for use
with large housings.
Trademark. Noryl
Supplier. General Electric Company

Properties	Impact modified	30% Glass filled
Flexural modulus, 10^3 psi @ 73°F	325-345	1,100-1,150
Mold linear shrinkage, in./in.	0.006	0.001-0.004
Tensile strength at break, psi	7000-8000	17,000-18,500
Tensile modulus, 10^3 psi	345-360	1200
Specific gravity	1.27-1.36	1.08-1.09
Hardness: Rockwell	L108, M93	R115-116
Water absorption, % (1/8 in. thick speci-men), 24 hr	0.1-0.07	0.06

Source: Ref. 7.

STYRENICS

Styrene is a component of acrylonitrile-butadiene-styrene (ABS) and styrene-acrylonitrile (SAN) copolymers, as well as polystyrene homopolymers.

4.4.6 Polystyrene (PS)

Properties and Characteristics

Low price; rigid with good dimensional stability; easily surface decorated; limited solvent resistance; resists mild acids and alkalis but attacked by solvents; limited heat resistance to approximately 160°F (71°C) unless using a heat-resistant grade; poor ultraviolet stability; low moisture absorption. Polystyrene is manufactured in a number of grades, including flame-retardant.

Fastening characteristics
   Torque. Attainable torques are in the medium-to-high range.
   Tensile. PS has the capability of accepting relatively high clamp
      loads. In unfilled grades, the assembly strength seldom reaches
      the level of fastener strength.
   Cracking. The fastening site requires moderate-to-large boss di-
      ameter to pilot-hole diameter ratios and strong boss support.
   Material relaxation. Only a slight loss of initial clamp load is ex-
      pected in the unfilled grades, even less will be found with filled
      materials.

If not used with excessive fillers (or in foam form), polystyrene is amenable to special-design thread-rolling screws for plastics and satisfactory for push-in thread-forming fasteners especially designed for plastics. Ultrasonic installation of internally threaded metal inserts should give good results in either the general purpose or impact grades and is satisfactory for use with polystyrene foam.

Applications Toys, novelties, appliance housings and liners, com-
      puter housings, television cabinets, impact-type furniture, auto-
      mobile parts, camera cases, luggage, lighting panels, cassettes,
      diagnostic equipment, etc.
   Tradenames. Dylene, Fosta Tuf-Flex, Lustrex, Styren, etc.
   Representative suppliers. Amoco, Arco, Dow, Gulf, Monsanto, Shell,
      American Hoechst, etc.

Properties	Homopolymer			
	Medium/high flow	Rubber modified	Heat resistant	20% Glass filled
Flexural modulus,	380-490	260-360	450-500	950-1000

| | Homopolymer | | | |
Properties	Medium/high flow	Rubber modified	Heat resistant	20% Glass filled
10^3 psi @ 73°F				
Mold linear shrinkage, in./in.	0.004-0.007	0.004-0.007	0.004-0.007	0.001-0.003
Tensile strength at break, psi	5200-7500	2700-6200	7100-8200	10,000-12,000
Tensile modulus, 10^3 psi	330-475	240-370	450-485	900-1100
Specific gravity	1.04-1.05	1.03-1.06	1.04-1.05	1.20
Hardness: Rockwell	M60-75	L50-82	M75-78	M80-95
Water absorption, % (1/8 in. thick specimen):				
24 hr	0.01-0.03	0.05-0.07	0.01	0.01
saturation	0.01-0.03	—	—	—

Properties	Flame-retardant (UL-40) rubber modified
Flexural modulus, 10^3 psi @ 73°F	320-330
Mold linear shrinkage, in/in	0.003-0.006
Tensile strength at break, psi	2650-4100
Tensile modulus, 10^3 psi	240-300

Properties	Flame-retardant (UL-40) rubber modified
Specific gravity	1.16-1.17
Hardness: Rockwell	L38-65
Water absorption, % (1/8 in. thick specimen), 24 hr	0.0

Source: Ref. 7

4.4.7 Acrylonitrile-Butadiene-Styrene (ABS)

Properties and Characteristics

Good chemical resistance being generally resistant to strong acids and alkalis; poor resistance to solvents and ultraviolet light; generally slightly hygroscopic; can be formulated or modified to meet requirements for flexibility or high rigidity, flame retardance, platability, or for improved weatherability; heat deflection temperature 230°F (110°C) reported available.

Fastening characteristics
Torque. With ABS, the greater the rubber (butadiene) content, the lower the torque values that will be achieved.
Tensile. Varies in a similar way, depending on rubber content.
Cracking. Varies, depending on rubber content. With low rubber content there is a tendency toward crazing on screw entry. This can be controlled by counterboring at entry site.
Material relaxation. Medium-impact grades have better resistance.

ABS works very well in unmodified, unfilled grades with special-design thread-rolling screws for plastics and should be excellent with push-in thread-forming fasteners especially designed for plastics. Ultrasonic insertion of internally threaded metal inserts gives excellent results in ABS because of the low viscosity at the melt temperature. With PVC-modified ABS, ultrasonic insertion is satisfactory.

Applications. Suitcases, tote boxes, appliance and power-tool housings, telephone cases, automotive cluster kick panels, grill work, headlight housings, crashpad and armrest covers, office machine and computer components and housings, safety equipment, etc.
Tradenames. Abson, Cycolac, Kralastic, Lustran, Royalex, etc.
Representative suppliers. Borg-Warner, Dow, Mobay, Monsanto, Uniroyal, etc.

		Injection grades		
ABS Properties	Extrusion grades	Heat resistant	Medium impact	High impact
Flexural modulus, 10^3 psi @ 73°F	130-420	300-400	350-400	250-350
Mold linear shrinkage, in./in.	—	0.004-0.009	0.004-0.009	0.004-0.009
Tensile strength at break, psi	2500-8000	6000-7500	6000-7500	4800-6300
Tensile modulus, 10^3 psi	130-380	300-350	300-400	230-330
Special gravity	1.02-1.06	1.05-1.08	1.03-1.06	1.01-1.04
Hardness: Rockwell	R75-115	R100-115	R107-115	R85-105
Water absorption, % (1/8 in. thick specimen), 24 hr	0.02-0.45	0.02-0.45	0.02-0.45	0.02-0.45

			Injection molding	
	Flame-retardant			20% Glass filled
Properties	ABS	ABS/PVC	Platable	
Flexural modulus, 10^3 psi @ 73°F	300-400	340	340-390	710
Mold linear shrinkage, in./in.	0.004-0.008	0.003-0.005	0.005-0.008	0.002

| | Flame-retardant | | | Injection molding |
ABS Properties	ABS	ABS/PVC	Platable	20% Glass filled
Tensile strength at break, psi	5000-8000	5800	6000-6400	11,000
Tensile modulus, 10^3 psi	320-400	330	320-400	740
Specific gravity	1.16-1.21	1.21	1.06-1.07	1.22
Hardness: Rockwell	R100-120	R100	R103-109	M85
Water absorption, % (1/8 in. thick specimen), 24 hr	0.2-0.6	—	—	—

Source: Ref. 7.

4.4.8   Styrene-Acrylonitrile Copolymer (SAN)

Properties and Characteristics

Improved chemical resistance over polystyrene (similar to ABS); significantly better ultraviolet resistance than ABS is available by addition of stabilizers; poor solvent resistance; slightly hygroscopic.

Fastening Characterisitcs
   Torque and tensile. Relatively high torque and tensile capabilities exist with SAN.
   Cracking. Possessing considerable rigidity, SAN has a tendency toward crazing on screw entry. This tendency is controllable by counterboring at entry site.
   Material relaxation. Assembly load relaxation is not severe.

SAN works very well in unfilled grades with special-design thread-rolling screws for plastics and should be excellent with push-in thread-forming fasteners especially designed for plastics. Ultrasonic installation of internally threaded metal inserts gives excellent results in SAN because of the low viscosity at melt temperature.

Applications. Automotive dashboard components, support panels, medical instruments, etc.

Tradenames. Acrylafil, Lustran, etc.

Representative suppliers. Borg-Warner, Dow, Monsanto, Uniroyal, etc.

		SAN Copolymers		
			High heat resistant	
Properties	Unfilled	20% Long Glass filled	Unfilled	20% Glass filled
Flexural modulus, 10^3 psi @ 73°F	500-580	1100-1280	450-490	800-870
Mold linear shrinkage, in./in.	0.003-0.005	0.001-0.003	0.005	0.003
Tensile strength at break, psi	10,000-11,900	15,500-16,000	7100-8100	10,700-12,000
Tensile modulus, 10^3 psi	475-560	1100-1280	440-490	850-900
Specific gravity	1.07-1.08	1.22-1.40	1.07-1.10	1.21-1.22
Hardness: Rockwell	M80-R83	M89-100	—	—
Water absorption, % (1/8 in. thick specimen):				
24 hr	0.15-0.25	0.01-0.2	0.1	0.1
Saturation	0.5	0.7	—	—

Source: Ref. 7.

4.4.9   Polyethersulfone (PES)

Properties and Characteristics

Strongest and with highest chemical resistance to acids, oils, gasoline, and concentrated alkalis of all amorphous thermoplastics; platable; excellent electrical properties; mildly hygroscopic; unpigmented product should not be used in exterior applications; available in grades that are usable up to 420°F (216°C); relatively high cost.

Fastening characteristics

Torque. Tests show quite high-torque and fail-torque values will be achieved with special-design thread-rolling screws for plastics.

Tensile. Exceptionally high pull-out values were achieved in the tests on PES as against other amorphous plastics.

Cracking. Reported to be a tough material even under fairly high stress and with good resistance to crack initiation. Absorption of moisture will increase toughness.

Material relaxation. PES has very low creep. Elastic recovery is reported superior to most engineering thermoplastics.

PES is highly amenable to the use of special-design thread-rolling screws for plastics. Because of the relatively high torque values encountered, prior testing of applications by a competent fastener laboratory is recommended to determine proper pilot hole and boss size. Not recommended for push-in thread-forming fasteners unless installed ultrasonically. We expect PES to have good characteristics for ultrasonic installation of internally threaded metal inserts. Since PES is available in very high, heat-resisting grades, prior testing is required to obtain the best unit settings.

Applications. Medical accessories, aerospace components, automotive and electrical parts with high heat requirements, business machines with thin-wall housing, pumps, valves, hot water meters.

Tradename. Victrex

Supplier. ICI Americas, Inc. (Imperial Chemical Industries Plc).

Properties[a]	PES	20% Glass filled	30% Glass filled
Flexural modulus 10^3 psi @ 68°F	373	856	1219
Mold linear shrinkage, in./in.	0.006	0.003	0.002

Properties[a]	PES	20% Glass filled	30% Glass filled
Tensile strength, psi @ 68°F	—	18,000	20,333
Tensile modulus, 10^3 psi	350	—	—
Specific gravity	1.37	1.51	1.60
Hardness: Rockwell	M88	M98	M98
Water absorption, % (1/8 in. thick speci- men), 24 hr	0.43	—	—

[a]Data from ICI Americas.

4.4.10 Polysulfone

Properties and Characteristics

Transparent; resists heat, heat deflection temperature 345°F (174°C); good resistance to chemical attack against acids, alkalis, oil, alcohol; good electrical properties; low water absorption; tough. Not recommended for exterior use without applied coating.

Fastening characteristics
  Torque. Polysulfone has relatively high drive and fail torque
    levels.
  Tensile. Tensile (pull-out) level can be expected to be relatively
    high.
  Cracking. Tends to be notch-sensitive, and fastening-site design
    should take sensitivity into consideration. We recommend that a
    competent fastener-testing laboratory be consulted on all appli-
    cations.
  Material relaxation. Very low, initial load relaxation can be ex-
    pected.

Amenable to special-design thread-rolling screws for plastics, under the preceding context. Not recommended for push-in thread-forming fasteners. Ultrasonic installation of internally threaded metal inserts may be recommended after testing.

  Applications. Used on certain under-the-hood automotive components,
    appliance parts, heat-fusing copy machines, medical accessories,
    television components, etc.

Trademark. Udel
Supplier. Union Carbide

Properties	Unfilled	Mineral filled	30% Glass filled
Flexural modulus 10^3 psi @ 73°F	390	600-750	1050
Mold linear shrinkage, in./in.	0.007	0.004-0.005	0.001-0.003
Tensile strength at break (psi)	—	9500-9800	14,500
Tensile modulus $(10^3$ psi)	360	550-650	1350
Specific gravity	1.24-1.25	1.48-1.61	1.46
Hardness: Rockwell	M69	M70-74	M90-100
Water absorption % (1/8 in. thick specimen):			
24 hr	0.3	—	0.3
Saturation	0.7	0.5-0.6	—

Source: Ref. 7.

4.5   THERMOPLASTIC STRUCTURAL FOAM

Properties and Fastening Characteristics

Structural foam is a relatively resilient material owing to its thin
skin and cellular structure.   It displays excellent thread-forming
abilities and is well suited for the use of special-design thread-
rolling screws for plastics which provide maximum joint strength versus standard thread-forming fasteners.

Limited tests reveal that the push-in thread-forming fastener for
plastics has potential use in structural foam.   However, the use of
thread-cutting screws is inadvisable owing to the cellular structure
of the material.   They also cut through the formed "skin" on the surface of the pilot hole and weaken joint structure.   Low overall assembly
strength results.   Structural foam is used in both large and small
structural applications which require rigidity, lightweight design, and
good mechanical properties.

The structural foam process is similar to injection molding except
that the resin is either mixed with an inert gas at the nozzle or premixed

with a chemical blowing agent. When heated, a gas is released which is dispersed through the polymer melt. Expansion of the melt permits density reduction and low mold pressure. This process creates an internal structure of random porosity around which a "skin" is formed. The skin formation provides greater density and, consequently, greater strength to the material's surface. When manufactured from a thermoplastic polymer, the structural foam material can be reprocessed.

> *Representative suppliers.* American Hoechst, Celanese, Dow, General Electric, Hercules, etc.

4.6 THERMOSET PLASTICS

4.6.1 Diallyl Phthalate (DAP)

Properties and Characteristics

Outstanding dimensional stability and electrical/electronic properties, (solder does not stick to it and material does not "outgas" in vacuum); flame-retardant; colorable.

> *Fastening characteristics*
> Torque. Moderate high drive and fail torques resulting in relatively low drive-to-fail ratios.
> Tensile. Displays reasonably high pull-out values.
> Cracking. DAP appears to have considerable resistance to cracking.
> Material relaxation. Creep is reported to be quite low. However, observed preliminary tests indicate unexpected creep levels. Further testing would be desirable.

Thread-cutting screws are generally recommended for thermoset plastics. The use of highly engineered, special-design thread-rolling screws for plastics may be considered for (at least) single insertion applications. Pilot holes should be counterbored. Consultation with a competent fastener laboratory is recommended for all applications.

Since there is virtually no postmold shrinkage, the use of molded-in inserts should be satisfactory in DAP. Not suitable for push-in thread-forming-type fasteners or ultrasonic installation of internally threaded metal inserts.

> *Applications.* Critical electrical and electronic switches, circuit breakers, standoffs, television components, circuit boards, bobbins in aerospace and computer applications, and with reinforcement: ducting, radomes, junction boxes, funiture laminates.
> *Tradenames.* Durez, Plaskon, Poly-DAP, Rogers, Sumikon
> *Representative suppliers.* Occidental (Hooker), Plaskon, Rogers

	DAP Molding compounds	
Properties	Glass filled	Mineral filled
Flexural modulus, 10^3 psi @ 73°F	1200-1500	1000-1400
Mold linear shrinkage, in./in.	0.0005-0.005	0.002-0.007
Tensile strength at break, psi	6000-11,000	5000-8000
Tensile modulus, 10^3 psi	1400-2200	1200-2200
Specific gravity	1.70-1.98	1.65-1.85
Hardness: Rockwell	E80-87	E61
Water absorption, % (1/8 in. thick specimen), 24 hr	0.12-0.35	0.2-0.5

Source: Ref. 7.

4.6.2 Amino Resins: Urea Formaldehyde and Melamine Formaldehyde

Properties and Characteristics

Amino resin polymers display heat resistance and have good electrical properties; like phenolics, however, they must be mixed with a filler to display hardness and strength; they are filled with wood flour, minerals, glass fiber, or chopped cotton flock.

Fastening characteristics
   Torque. When using thread-cutting screws, only low drive torques will be achieved. With any type of threaded fastener, cracking of fastening site is likely, unless seating torque is carefully selected.
   Tensile. Quite high tensile strength should be expected.
   Cracking. Susceptible to cracking, and in most cases thread-cutting screws should be used.
   Material relaxation. Virtually no relaxation.

   Type BT standard thread-cutting screws are generally recommended for thermoset plastics. Some minimal success has been achieved with special-design thread-rolling screws for plastics, and a competent fastener-testing laboratory may be consulted in this regard. Not suitable for push-in thread-forming fasteners or ultrasonic installation of internally threaded metal inserts.

Applications. Electrical switches, camera cases, appliance housings, etc.

Tradename. Durez

Representative suppliers. American Cyanamid, Borden, Georgia-Pacific, Gulf, Occidental, Monsanto, etc.

Properties	Urea formaldehyde alphacellulose filled
Flexural modulus, 10^3 psi @ 73°F	1300-1600
Mold linear shrinkage, in./in.	0.006-0.014
Tensile strength at break, psi	5500-13,000
Tensile modulus, 10^3 psi	1000-1500
Specific gravity	1.47-1.52
Hardness: Rockwell	M110-120
Water absorption, % (1/8 in. thick specimen), 24 hr	0.4-0.8

Source: Ref. 7.

4.6.3 Phenolic

Properties and Characteristics

Strong, rigid, and heat-resistant; dimensional thermal stability; low cost; excellent electrical properties and chemical resistance; scratch-resistant. However, phenolics must be modified with a filler to be usable as they are too brittle and weak by themselves. Phenolics are often filled up to 40–50% with wood flour, asbestos, or minerals. Glass fiber can also be added to improve impact resistance and toughness.

Fastening characteristics
   Torque. Low drive torque will be achieved, using thread-cutting screws. Select seating torque carefully to avoid cracking fastening site.

Tensile. Relatively high tensile strength.
Cracking. Very susceptible to cracking, and in most cases only
thread-cutting screws should be used.
Material relaxation. Virtually no relaxation.

Type BT standard thread-cutting screws are generally recommended
for thermosets. The internal thread quality is generally poor and often deteriorates further on multiple use. Special-design thread-rolling
screws for plastics may be recommended for single insertions when
tested by a competent fastener testing laboratory. Not suitable for
push-in thread-forming fasteners or ultrasonic installation of internally
threaded metal inserts.

Applications. Steam iron handles, toaster parts, housings, washing
machine agitators, pump impellors, automotive ignition parts, electrical switch gear, etc.
Trademarks*. Bakelite, Durez, Genal, Ucar, etc.
Representative suppliers*, Occidental, Union Carbide and former
international subsidiaries, etc.

| | Impact modified | | | |
Properties	Cotton filled	Cellulose filled	Fabric and rag filled	Wood flour filled
Flexural modulus, 10^3 psi @ 73°F	800-1300	900-1300	900-1300	1000-1200
Mold linear shrinkage, in./in.	0.004-0.009	0.004-0.009	0.004-0.009	0.004-0.009
Tensile strength at break, psi	6000-10,000	3500-6500	6000-8000	5000-9000
Tensile modulus, 10^3 psi	1100-1400	—	900-1100	800-1700

*Note. There have been a number of recent changes in phenolic prod-
uct lines. See Chap. 5.

	Impact modified			
Properties	Cotton filled	Cellulose filled	Fabric and rag filled	Wood flour filled
Specific gravity	1.38-1.42	1.38-1.42	1.37-1.45	1.37-1.46
Hardness: Rockwell	M105-120	M95-115	M105-115	M100-115
Water absorption, % (1/8 in. thick specimen), 24 hr	0.6-0.9	0.5-0.9	0.6-0.8	0.3-1.2

Source: Ref. 7.

4.6.4 Thermoset Polyester: Unsaturated Polyester or Alkyd Polyester

Properties and Characteristics

The unsaturated polyester is almost always used with a high percentage of glass fiber filler. When used with structural processes for autobodies and like applications or components, it is compression molded and identified by various molding characteristics: bulk molding compound (BMC), dough molding compound (DMC), sheet molding compound (SMC), or thick molding compound (TMC). Besides being highly amenable to use with reinforcements, unsaturated polyesters are corrosive-resistant and have outstanding electrical properties.

Fastening characteristics
Torque. Thread-cutting screws will only achieve low torque levels. Where fastening site is designed properly, the special-design thread-rolling screws for plastics should produce relatively high values.
Tensile. Pull-out strength can surpass screw strength.
Cracking. Pilot holes should be counterbored, as material is subject to crazing at the entry hole.
Material relaxation. Virtually none.

Although type BT standard thread-cutting screws are generally recommended for thermoset plastics, we recommend testing with special-design thread-rolling screws for plastics, as there have been surprisingly successful uses. (See Chap. 5, References.) A competent fastener testing laboratory should be consulted.

Neither use of push-in thread-forming fasteners or ultrasonic installation of internally threaded metal inserts is recommended. Molded-in
metal inserts require substantial amounts of surrounding material (6).

Applications. (A) Reinforced polyester: sports car bodies, electrical appliance and business equipment housing, bathtubs, shower
stalls and sanitary ware, automotive exterior components, chemical
tanks, hoods, ducts, decorative panels, motor housings, hand
tools, etc. (B) Alkyd polyester: automotive ignition and elec-
trical/electronic applications.
Tradenames. Glaskyd, Glastic
Suppliers. American Cyanamid, Glastic

Properties	Reinforced unsaturated	Alkyd polyester mineral filled
Flexural modulus, 10^3 psi @ 73°F	1000-2000	2000
Mold linear shrinkage, in./in.	0.001-0.012	0.003-0.010
Tensile strength at break, psi	3000-10,000	3000-9000
Tensile modulus, 10^3 psi	1000-2500	500-3000
Specific gravity	1.65-2.30	1.6-2.3
Hardness: Rockwell	50-80 Barcol	E98
Water absorption, % (1/8 in. thick speci- men), 24 hr	0.06-0.28	0.05-0.5

Source: Ref. 7.

4.7   THERMOSET FOAM PLASTICS

4.7.1   Urethane Foam

Properties and Characteristics

Thermoplastic polyurethanes are available in elastomeric formulations
with widely varying characteristics. In their foam form they can be
quite flexible and soft or spongelike. Under those circumstances the
material is not amenable to the use of thread-rolling screws. Perform-
ance-testing in the spongelike urethane foams revealed exceedingly

low installation and strip torques for all test instances. Ultrasonic
insertions are not recommended.

4.7.2   Reaction Injection Molding (Reinforced Reaction Injection Molding) Process

Properties and Characteristics

The reaction injection molding RIM process uses two-component polymers, such as urethane, which are placed in separate displacement
piston-metering cylinders and injected directly into the mold at high
pressure through a mixing nozzle. Sometimes high-capacity metering
pumps are used. After mixing, the two components undergo a heat-
producing (exothermic) reaction in the mold, requiring little or no
added heat or pressure.

Reinforcements such as milled glass, wollastonite, or similar materials can be added to enhance the physical characteristics of the cured
product.

Since RIM or reinforcement reaction injection molding (RRIM) is a
process, other two-component materials could be substituted for urethane, which is the primary, current choice. Epoxies or polyesters
are the current alternatives. The epoxies, at least, are quite expensive. Also, there are reports on the use of Nylon for the RIM process.

Advantages include: significantly lower (10–30% energy requirement), reduced cost of molding from lower pressure requirements, design flexibility, ribbing, and bosses easier to fill, and, short cycle
time (currently 2-4 min).

Mold cleaning, however, may represent an increased difficulty in the
more complex shapes.

Fastener characteristics: In connection with fasteners, the biggest
advantage of the RIM process is that no stresses are created in the
cured product, during the molding operation. Provided the material
has sufficient rigidity, fastening characteristics may prove satisfactory.
Testing by a competent fastener laboratory is recommended.

TRADEMARKS CROSS-REFERENCED IN CHAPTER 5

Bakelite	Glaskyd
Celanex	Glastic
Celcon	Kralastic
Cycolac	Lexan
Delrin	Merlon
Durez	Minlon
Gafite	Moplen
Gaftuf	Noryl
Genal	Nylafil

Polycarbafil	Tivar
Profax	Ultem
Profil	1900 UHMW Polymer
Rexene	Valox
Royalex	Vespel
Rynite	Victrex
Ryton	Vydene
Teflon	Zytel
Tenite	

REFERENCES

1. *Machine Design*, 1982 materials reference issue 54:147.

2. J. H. Crate, *Plastics Design Forum* 6:61-64.

3. R. E. Schultz, *Plastics Design Forum* 6:34 (March/April 1981).

4. R. J. Kahl, *Plastics Products Design Handbook* (E. Miller, Ed.), Marcel Dekker, New York, 1981, p. 231.

5. *Ultem Resin*, General Electric Publication, Table 14, pg. 73.

6. *Plastics Design Forum* 7:91 (May/June 1982).

7. J. Agranoff, Ed., *Modern Plastics Encyclopedia*, Vol. 58, No. 10A, McGraw-Hill, New York, 1981-1982.

BIBLIOGRAPHY

Agranoff, J. H., Ed., *Modern Plastics Encyclopedia*, Vol. 57, No. 10A, Vol. 58, No. 10A, McGraw-Hill, New York, 1981-1982.

Characteristics of Thermoplastics for Ultrasonic Assembly Applications, Technical Bulletin 8101, Sonics & Materials, Inc., Danbury, Conn., 1981.

The Handbook of Engineering Structural Foams, General Electric, SFR-3, January 1978.

Polyethylene terephthalate, *Plastics Design Forum* (January/February) 1982).

5
TRADENAME PLASTICS

It will be noted that many companies use only one tradename to cover a variety of materials. Bakelite is an example. Bakelite has been the tradename for all plastics manufactured by Union Carbide and its former, and now divested, international subsidiaries, but it was often mistakenly characterized as referring only to the grades of phenolics produced by that organization.

We should caution that, as noted elsewhere in this volume, the quantity of fillers (such as glass or mineral) in plastic materials, as well as the blend of resins, greatly affects the performance and choice of fasteners. Since the percentage of fillers and resin blends can vary widely, a competent fastener-testing laboratory or the resin producer should be consulted about specific materials.

In this chapter we do not mention all the plastics covered under their generic names in the previous chapter. This does not mean that a particular generic name plastic has not been tested.

5.1 BAKELITE®–PHENOLIC (UNION CARBIDE CORPORATION, SPECIALTY CHEMICALS AND PLASTICS DIVISION, New York, NY)

Union Carbide is now in the process of gradually changing over the Bakelite tradename to a new one (UCAR) that identifies its products more closely with the corporate name. The former subsidiaries have retained the older trademark usage.

We tested Bakelite as a phenolic (Chap. 4, Sec. 4.6.3). In this form Bakelite is a brittle thermoset material displaying little ductility and is not generally amenable to thread-rolling screw principles, especially when compression-molded. Internal thread quality is generally poor and often deteriorates on multiple reuse.

However, applications which require single installation can often
successfully use special-design thread-rolling screws for plastics. If
repeated field-assembly operations are anticipated, thread-rolling
screws are generally not recommended in this material. Use of push-
in thread-forming fasteners for plastics is not desirable in phenolics.
In many cases, thread-cutting screws provide adequate performance.
Ultrasonic insertion is not recommended in thermosets.

5.2 CELANEX® (CELANESE PLASTICS AND SPECIALTIES COMPANY, Chatham, NJ)

A thermoplastic polyethylene terephthalate (PET) polyester (see
Chap. 4, Sec. 4.2.6) which is characterized as being hard, strong,
and extremely tough with low creep. Celanex is available in a variety
of grades including reinforcements with mineral, glass, or both and is
highly adaptable to automotive, electrical, and houseware applications.
    Our tests with this material some time ago, using an early special-
design thread-rolling screw for plastic showed rather satisfactory
drive-to-strip ratios ranging up to approximately 5:1 with high differential.
    Although we would not recommend use of standard thread-forming
screws in this material, we are confident that the more highly engineered special-design thread-rolling screws for plastics would produce
optimum thread-forming performance and could provide satisfactory
assembly strength when used in the proper hole size and with careful
fastening-site design. Thread-cutting screws provide adequate performance in nondemanding applications.
    Ultrasonic installation of internally threaded metal inserts and push-
in-type thread-forming fasteners is recommended.

5.3 CELCON® (CELANESE PLASTICS AND SPECIALTIES COMPANY, Chatham, NJ)

A copolymer, acetal-based engineering resin (Chap. 4, Sec. 4.2.1)
which displays excellent thread-rolling and thread-forming characteristics, Celcon is a material recommended for use with both special-
design thread-rolling screws and push-in thread-forming fasteners
for plastics. It is also recommended for ultrasonic installation of internally threaded metal inserts.

5.4 CYCOLAC® (MARBON DIVISION, BORG-WARNER CHEMICALS, Parkersburg, WV)

An acrylonitrile-butadiene-styrene (ABS) resin (Chap. 4, Sec. 4.4.7)
which is characterized by excellent cold-forming and cold-rolling

properties. The exceptional resiliency of this material allows for successful use of thread-forming and special-design thread-rolling screws for plastics over a wide range of hole sizes for any given diameter. Push-in thread-forming fasteners for plastics should work well in Cycolac. ABS resins are excellent materials for ultrasonic installation of internally threaded metal inserts.

5.5 DELRIN® (E.I. DU PONT DE NEMOURS AND CO., Wilmington, DE)

This is a homopolymeric (single polymer), acetal-based resin (Chap. 4, Sec. 4.2.1) which offers excellent thread-forming and thread-rolling abilities. The material is characterized by low-drive or installation torque and high-failure torque. It also offers high tensile pull-out strengths. Generally, Delrin is an excellent material for either special-design thread-rolling screws or push-in thread-forming fasteners for plastic. It should give good results with ultrasonic insertions.

5.6 DUREZ® (OCCIDENTAL CHEMICAL CORPORATION, North Tonawanda, NY)

Durez is the tradename of a family of thermoset phenolics, alkyds, and diallyl phthalates (DAP) produced by Occidental Chemical Corporation. Thread-cutting screws are usually recommended for these materials.

As a phenolic resin (Chap. 4, Sec. 4.6.3), Durez usually displays slight ductility and is neither generally suitable for use in thread-forming by special-design thread-rolling screws or push-in thread-forming fasteners for plastic. Its brittleness often results in poor internal thread quality and prevents multiple removal cycles.

However, several injection-molded grades demonstrate compatibility and should work well with the newer types of special-design thread-rolling screws for plastics.

Durez diallyl phthalate (DAP) (Chap. 4, Sec. 4.11) appears to display some resistance to cracking and has some promise of compatibility with the more highly engineered special-design thread-rolling screws for plastics, at least for single insertions.

The fastener manufacturer's product-engineering department should be consulted on specific applications for Durez materials. Ultrasonic insertions are not recommended with thermosets.

5.7 FIBERESIN® (FORMERLY MADE BY UNITED STATES GYPSUM)

Fiberesin was a composite material of wood fibers and thermosetting plastic resin, probably phenolic (See Chap. 4, Sec. 4.6.3) which was produced as a reinforced, solid plastic panel. This type of material is primarily used by the furniture industry to fabricate decorative and structural office furniture.

The material displays some brittleness (mainly demonstrated through surface crazing) and a powdery drilled chip which could limit effective threaded fastener use in thin engagements.

The material is not suitable for the use of push-in thread-forming fasteners for plastics, or for ultrasonic installations of internally threaded metal inserts.

Certain special-design thread-rolling screws for plastic seem to provide a reasonable balance of performance efficiency and fastening costs in Fiberesin products. For this reason, we recommend submitting Fiberesin-type applications for evaluation by the fastener manufacturer's product engineering department, in all instances.

5.8   GAFITE®–GAFTUF® (GAF CORPORATION, ENGINEERING PLASTICS DEPARTMENT, New York, NY)

Gafite is a thermoplastic polyester (See Chap. 4, Sec. 4.2.6) which is also available in a more ductile grade called Gaftuf. This material comes in unfilled grades as well as in grades filled with glass and/or mineral.

Our extensive testing indicates that special-design, trilobular roll-thread screws for plastics provide excellent assembly strength and good drive-to-strip ratios in Gafite applications. Push-in thread-forming fasteners for plastics should be functional in the ductile grades. It has good characteristics for ultrasonic installation of internally threaded metal inserts.

5.9   GENAL®* (GENERAL ELECTRIC COMPANY, SPECIALTY PLASTICS DIVISION, Pittsfield, MA)

An injection and compression moldable phenolic thermoset (See Chap. 4, Sec. 4.6.3), Genal was available in a variety of grades. Although thermoset phenolics are not generally good thread-rolling or thread-forming materials, some limited success with special-design thread-rolling screws for plastics was achieved with this material. Thread-cutting screws generally provide adequate performance in phenolics.

The fastener manufacturer's product engineering department should be consulted for all applications involving this kind of material, as new designs of special screws for plastic may be more suitable. Push-in

---

*It is our understanding that the Genal line of phenolics was discontinued by General Electric in December, 1982.

thread-forming fasteners for plastics should not be used with phenolics. Ultrasonic insertion is not recommended.

5.10 GLASKYD® (AMERICAN CYANIMID COMPANY, ORGANIC CHEMICAL DIVISION, Bound Brook, NJ)

This is a thermoset, alkyd-molding compound (see Chap. 4, Sec. 4.6.4) which shows some brittleness and a very powdery drilled chip. However, among the variety of grades (surprisingly) good thread-forming ability will be found when using certain special-design thread-rolling screws for plastic.

When possible, it is important to counterbore the pilot holes in Glaskyd applications to prevent surface crazing.

Standard thermoset hole-size recommendations for these special-design screws are generally satisfactory. See manufacturers' catalogs.

It is not a suitable material for use with push-in thread-forming fasteners or for ultrasonic installations.

5.11 GLASTIC® (GLASTIC CORPORATION, Cleveland, OH)

Glastic is a glass-fiber-reinforced thermoset polyester (see Chap. 4, Sec. 4.6.4) which displays excellent thread-forming ability and is well suited for the use of special-design thread-rolling screws for plastics. Optimum pilot hole sizes for this material generally conform to the thermoset recommendations. It is not a suitable material for use with push-in thread-forming fasteners for plastics or ultrasonic insertions.

5.12 KRALASTIC® (UNIROYAL CHEMICAL, DIVISION OF UNIROYAL, INCORPORATED, Naugatuck, CT)

This is a basic ABS plastic (see Chap. 4, Sec. 4.4.7) which is an excellent material for both special-design thread-rolling screws and push-in thread-forming fasteners for plastics. It is also an excellent material for ultrasonic installation of internally threaded metal inserts. ABS products are exceptionally versatile plastics that combine, in a single material, toughness and chemical resistance, and they can be modified for various degrees of hardness of rigidity. Owing to these characteristics. ABS products often replace metal parts.

5.13 LEXAN® (GENERAL ELECTRIC COMPANY, SPECIALTY PLASTICS DIVISION, Pittsfield, MA)

This material is an exceedingly tough polycarbonate resin (see Chap. 4, Sec. 4.4.2) which displays somewhat limited thread-forming capabilities.

5.13.1 Standard Screws

Quantitative evaluation has shown that 1500—2000 psi of hoop stress are introduced when standard thread-forming screws are used with Lexan polycarbonate. Consequently, the resin producer recommends standard thread-cutting screws which cause stresses of approximately 1000 psi.

5.13.2 Special-Design Screws for Plastic

Lexan is a high-stress, concentrated material, a factor which renders hole-size recommendations a critical parameter in many applications. Research is currently being done with special-design thread-rolling screws to establish hole size and boss outside diameter relationships that could result in an acceptable stress level. There have been a number of field successes in Lexan using the more highly engineered special-design thread-rolling screws for plastics.

Lexan parts are frequently designed with thin-wall bosses, offering minimal strength for thread-forming. For this reason, the fastener manufacturer's product engineering department should be consulted for all Lexan applications.

Use of push-in thread-forming fasteners or push-in inserts for plastic are not recommended. However, ultrasonic installation of internally threaded metal inserts or push-in thread-forming fasteners (when ultrasonically installed) should prove satisfactory.

5.14 MERLON® (MOBAY CHEMICAL COMPANY, Louisville, KY)

Merlon products offered the initial use of polycarbonate resins (see Chap. 4, Sec. 4.4.2) and are found in a wide range of engineering plastics. It is characteristically a rigid and tough material which offers excellent thread-forming abilities. With suitable boss design, special-design thread-rolling screws for plastic should be readily usable. Ultrasonic installation of internally threaded metal inserts and push-in thread-forming fasteners should be good.

However, because of the nature of this material, Merlon applications are designed with thin-wall boss thicknesses. Therefore, the fasteners manufacturer's product engineering department should be consulted for all applications using this material.

5.15 MINLON® (E. I. DU PONT DE NEMOURS AND CO., Wilmington, DE)

This material is an engineering thermoplastic resin with a mineral-and-glass-reinforced polyamide base (see Chap. 4, Sec. 4.2.5). Our tests indicate it offers excellent drive-to-fail torque ratios and high tensile pull-out strengths.

Special-design thread-rolling screws for plastics are recommended in most Minlon applications. Push-in thread-forming fasteners for plastics may be considered, but the applications should be tested by the fastener manufacturer's product engineers. Ultrasonic installations should be satisfactory.

5.16 MOPLEN® (USS NOVAMONT, INC., Pittsburgh, PA)

Moplen is a polypropylene-based resin (see Chap. 4, Sec. 4.1.2) available in a variety of grades. This material is generally well suited for the use of both special-design thread-rolling screws and push-in thread-forming fasteners for plastics. However, because of the wide variety of available grades, the fastener manufacturer's product engineering department should be consulted for specific recommendations.

Polypropylenes are good materials for ultrasonic installations of internally threaded metal inserts; however, the holding power of polyolefins is not particularly high and insert design should consider this factor.

5.17 NORYL® (GENERAL ELECTRIC COMPANY, SPECIALTY PLASTICS DIVISION, Pittsfield, MA)

A modified phenylene oxide, engineering resin (see Chap. 4, Sec. 4.4.5) with excellent cold-forming abilities, Noryl is a thermoplastic of high resiliency, available in many grades. Noryl's elastic memory enhances the locking action inherent in special-design, trilobular, thread-rolling screws for plastic. It is an ideal material for special-design thread-rolling fasteners and push-in thread-forming fasteners for plastics.

The resin producer, however, has not endorsed the use of any screw other than a thread cutter. As with Lexan, it is our understanding that because of the reusability problem with thread-cutting screws and the successful use of special-design thread-rolling screws for plastics, the resin producer is reexamining this position, Noryl is suitable for ultrasonic installation of internally threaded metal inserts.

5.18 NYLAFIL® (FIBERFIL DIVISION, DART INDUSTRIES, INC., Evansville, IN)

A glass fiber-reinforced Nylon resin (see Chap. 4, Sec. 4.2.5), this material offers good assembly strength with special-design thread-rolling screws for plastic, provided sufficient engagement length is employed. Because of the reinforcement, Nylafil would not be recommended with push-in thread-forming fasteners for plastic. Ultrasonic insertion is recommended, however.

5.19 POLYCARBAFIL® (FIBERFIL DIVISION, DART
      INDUSTRIES, INC., Evansville, IN)

A glass fiber-reinforced polycarbonate resin (see Chap. 4, Sec. 4.4.2)
with excellent characteristics when used with special-design thread-
rolling screws for plastic. Ultrasonic installation of internally threaded
metal inserts and push-in thread-forming fasteners should be quite
satisfactory.

5.20 1900 ULTRAHIGH MOLECULAR WEIGHT POLYMER
      (HERCULES, INC., Wilmington, DE)

Polyethylene-based resins (see Chap. 4, Sec. 4.1.1) generally display
low torque and tensile pull-out strengths because of their exceedingly
elastic nature. The 1900 ultrahigh molecular weight (UHMW) polymer
is a high-density polyethylene resin which, provided there is sufficient
thread-engagement length, will offer good assembly strength with
special-design thread-rolling screws or push-in thread-forming fasteners for plastics. Ultrasonic installations are not recommended owing
to the high molecular weight.

5.21 PRO-FAX® (HERCULES, INC., Wilmington, DE)

Both Pro-Fax 6523 and (high density) 7523 are polypropylene-based
resins (see Chap. 4, Sec. 4.1.2) which are amenable to thread-rolling
and thread-forming principles. Special-design thread-rolling screws
and push-in threaded-fasteners for plastics are recommended.
    Although both plastics are resilient and offer good assembly strength
with thread-rolling screws, our tests indicate that the best assembly
strength can be obtained with the high-density Pro-Fax 7523.
    Ultrasonic installation should prove satisfactory in both grades. How-
ever, polypropylenes are not especially noted for their holding power
with fasteners, and it may be wise to consider this factor in selecting
the ultrasonic insert design.

5.22 PROFIL® (FIBERFIL DIVISION, DART
      INDUSTRIES, INC., Evansville, IN)

Profil is a glass fiber-reinforced polypropylene resin (see Chap. 4,
Sec. 4.1.2) with good thread-forming abilities. Test results with this
material indicate good assembly strength with both special-design
thread-rolling screws and push-in thread-forming fasteners for plastic. Ultrasonic installations are satisfactory, but note the remarks
about the previous material, which is also polypropylene based.

5.23 ROYALEX® (UNIROYAL CHEMICAL, DIVISION OF UNIROYAL, INCORPORATED, Naugatuck, CT)

An ABS sheet-material (see Chap. 4, Sec. 4.4.7) used in a wide variety of structural applications. In general, this material offers excellent assembly strength with both special-design thread-rolling screws and push-in thread-forming fasteners for plastics. Test results have shown this material to provide excellent drive-to-fail torque ratios and high tensile pull-out strengths. Ultrasonic installation of internally threaded metal inserts should prove very satisfactory.

5.24 RYNITE® (E. I. DU PONT DE NEMOURS AND CO., Wilmington, DE)

This is a thermoplastic polyester (PET) resin (see Chap. 4, Sec. 4.2.6), reinforced with glass or mica (or both) in several different grades, currently ranging from 30% filler content up to 55%. Rynite has good dimensional stability and is very stiff. For example, Rynite 555 has a flexural modulus of $2,600 \times 10^3$ psi and is claimed to be the stiffest glass-reinforced thermoplastic available. It is used in structural and electrical applications where mechanical fasteners are necessary.

Trilobular, special-design thread-rolling screws for plastics have been tested in several grades of Rynite, with good success. For optimum performance, pilot-hole size should coincide with recommendations normally given for thermoset plastics. It is expected that push-in thread-forming fasteners for plastics will function in some grades. Ultrasonic applications are recommended at less than 40% filler content only.

5.25 RYTON® (PHILLIPS CHEMICAL COMPANY, PLASTICS TECHNICAL CENTER, Bartlesville, OK)

Ryton is a thermoplastic polyphenylene sulfide resin (see Chap. 4, Sec. 4.2.7) available in a variety of glass fiber-reinforced and unreinforced grades for injection and compression molded parts.

The material is generally characterized by having a hard, rigid structure with both high tensile and flexural strength. Although a thermoplastic, this material displays the thread-rolling characteristics of some thermosetting resins. Most notable is the development of crazing on the material's surface. This crazing can present both strength and cosmetic problems in some applications, but can be somewhat alleviated by counterboring the pilot holes.

The material also shows evidence of a susceptibility to stress cracking. On the other hand, special-design thread-rolling screws for plastic can provide good performance under proper conditions. For this reason, the fastener manufacturer's product engineering department should be consulted for each Ryton application. The use of push-in thread-forming fasteners for plastics is not recommended unless inserted ultrasonically. Ultrasonic insertion of internally threaded metal inserts is recommended.

5.26   TEFLON® (E. I. DU PONT DE NEMOURS AND CO., Wilmington, DE)

This is a popular tetrafluoroethylene (TFE) and fluoronated ethylene propylene (FEP) fluorocarbon resin (see Chap. 4, Sec. 4.2.4) which is available for both coatings or molded products, Teflon is exceedingly resilient and lubricious, which permits successful thread formation in a wide range of hole sizes. However, these properties also lend themselves to very low drive and strip torques, possibly resulting in difficulty with drive-tool adjustment. We recommend consideration of filled product if the assembly encompasses threaded fasteners and use of those screws characterized as providing prevailing torque, such as the special-design trilobular screws for plastics.

Careful examination of all Teflon applications should be made before recommending hole sizes. Deep thread engagements are recommended for this material. Teflon is not recommended for ultrasonic insertion or mechanical insertion of push-in fasteners.

5.27   TENITE® (EASTMAN CHEMICAL PRODUCTS, INC., PLASTICS DIVISION, Kingsport, TN)

5.27.1   Tenite-Polyethylene

Polyethylene-based resins (see Chap. 4, Sec. 4.1.1) generally display low torque and tensile pull-out strengths because of the elastic nature of the resin. (See also remarks about ultrasonic assemblies in Chap. 4.)

5.27.2   Tenite-Polypropylene

Polypropylene (see Chap. 4, Sec. 4.1.2) ordinarily provides excellent performance and assembly characteristics with thread-forming and special-design thread-rolling screws and push-in thread-forming fasteners for plastics. It is available in a variety of general purpose and special formulations, including filled grades. Ultrasonic installations should prove satisfactory. However, polypropylenes are not noted for extraordinary holding power and the design of the ultrasonic insert should be selected accordingly.

Because of the variety of available grades for both of these Tenite plastics, the fastener manufacturer's product engineers should be consulted.

5.27.3   Tenite-Cellulose Proprionate

Cellulose proprionate (CAP) (see Chap. 4, Sec. 4.2.3) provides relatively high torque and tensile capabilities. To avoid stress cracking, slightly enlarged hole dimensions are recommended. Otherwise, it should prove highly amenable to special-design thread-rolling screws for plastics and push-in fasteners. Ultrasonic installations are recommended.

5.28 TIVAR® (POLY-HI/MENASHA CORP., Fort Wayne, IN)

Tivar is an ultrahigh-molecular-weight polymer. It is a polyethylene-based resin (see Chap. 4, Sec. 4.1.1) available in a variety of grades which offer excellent abrasion and corrosion-resistant qualities.

Previous experience has shown us that polyethylene resins often display low torque and tensile pull-out strength because they are exceedingly resilient. However, Tivar is a high-density polyethylene which, provided there is sufficient thread-engagement length, could offer good assembly strength with special-design thread-rolling screws and push-in thread-forming fasteners for plastics. Ultrasonic insertion is not recommended in UHMW materials.

Owing to the wide variety of Tivar grades and products, specific recommendations should be made by the fastener manufacturer's product engineering department for individual applications.

Performance tests have illustrated that trilobular, special-design screws for plastic offer excellent thread-forming capabilities in Tivar.

5.29 ULTEM® (GENERAL ELECTRIC COMPANY, SPECIALTY PLASTICS DIVISION, Pittsfield, MA)

Ultem is a newly developed polyetherimide which we have classified among the high-performance, amorphous thermoplastics (see Chap. 4, Sec. 4.4.3) and which exhibits a broad range of highly superior characteristics.

The product is available unfilled or glass filled and because of its interesting characteristics, combining elasticity and structural strength it is highly amenable to the use of special-design thread-rolling screws for plastic with at least up to 10% glass content.

In these grades, ultrasonic insertion is also recommended and the use of push-in fasteners for plastic should be considered. Because of the relative newness of Ultem to the assembly scene we suggest that a competent fastener laboratory be consulted.

5.30 VALOX® (GENERAL ELECTRIC COMPANY, SPECIALTY PLASTICS DIVISION, Pittsfield, MA)

Valox is a thermoplastic polyester resin (see Chap. 4, Sec. 4.2.6) which is the basis for a number of product variations including impact modification, glass, mineral, and glass-mineral reinforcements, and glass-resin alloys. The adaptability of this resin family allows it to be used in diversified markets such as electrical, automotive, houseware, fluid, and material handling.

These resins have proved satisfactory for use with special-design thread-rolling screw profiles and the push-in threaded fasteners developed for use in plastics. Ultrasonic installations should prove very satisfactory.

5.31 VESPEL® (E. I. DU PONT DE NEMOURS AND CO., Wilmington, DE)

A unique polyimide that is classified as a thermoplastic but acts like a thermoset in that it has no melting point. It also has outstanding electrical properties. Applications include such high-reliability usages as space hardware, aircraft components, and military hardware.

A minor but interesting usage for Vespel is as insert-rings in lock nuts, where it functions like Nylon but has very high temperature capability. The material has the ability, when molded as screws, to create its own thread in pilot holes formed in soft metals such as aluminum. We understand (from the manufacturer) that it has been used for manufacturing thread-forming screws for use where both extra electrical insulation and high-temperature ability are required and where the high cost is justified against the lower cost of mass produced metal thread-forming screws.

Vespel has good ability to accept the use of special-design thread-rolling screws for plastics. (See Chap. 4, Sec. 4.4.4 for further remarks.)

Ultrasonic insertions are not recommended.

5.32 VICTREX® (ICI AMERICAS, Wilmington, DE)

Victrex is a polyethersulfone developed by the Plastics Division of Imperial Chemical Industries PLC, Welwyn Garden City, Herts, England, and made available in the United States through ICI Americas. Owing to its high strength, chemical resistance, and usability up to 420°F (216°C), Victrex finds application in automotive, electrical, and aerospace components with high-heat requirements. The material appears very suitable for fastening applications using high-performance, special-design thread-rolling screws for plastic. Not recommended for use with axially installed push-in fasteners, except by ultrasonic means. Victrex should prove very suitable with ultrasonic installations of push-in fasteners or internally threaded metal inserts.

5.33 VYDYNE® (MONSANTO COMPANY, St. Louis, MO)

Vydyne is a mineral-reinforced polyamide (see Chap. 4, Sec. 4.2.5) which displays good thread-rolling characteristics. The most popular grade is Vydyne R, which is designed to replace nonferrous die castings. Our tests indicate that this grade is particularly well suited for special-design thread-rolling screws for plastics and ought to test out well with push-in threaded fasteners for plastics.

Ultrasonic insertion should be quite satisfactory provided the percentage of reinforcement is not too high.

5.34 ZYTEL® (E. I. DU PONT DE NEMOURS AND CO., Wilmington, DE)

A trademark for numerous Nylon resins (see Chap. 4, Sec. 4.2.5) manufactured by Du Pont, Zytel (Nylon) resins are thermoplastic polyamides available in both unreinforced and glass-reinforced versions. We have tested samples of both types and have achieved excellent results with special-design thread-rolling screws for plastic. In some grades, push-in thread-forming fasteners for plastics ought to be satisfactory. Ultrasonic installations have proved very satisfactory in Nylon, provided the amount of glass fill is not excessive.

Because of the variety of available grades of this material, the fastener manufacturer's product engineering department should be consulted for all applications.

6

FASTENING-SITE DESIGN
CONSIDERATIONS: THREADED FASTENERS

The first part of this chapter will cover fastening-site design for self-threading screws and provide suggestions for rectifying problem sites. The second part of the chapter will deal with the subject of tapping in plastics.

6.1 FASTENING-SITE DESIGN: SELF-THREADING SCREWS

Proper design of the fastening site plays a major part in the successful use of self-threading screws. Most fastening areas consist of protruding posts (Fig. 6.1) called bosses, which have taper-cored pilot holes. The bosses are extended at 90 degrees from a cross section and are sometimes rib-supported. When extra support is required, bosses are sometimes also attached to side walls.

6.1.1 Pilot Holes

Hole size is determined by examining a combination of screw performance requirements, such as torque and pullout, which result from the interplay of three major factors: (1) the size of the fastener, (2) the type of fastener, and (3) the plastic material used. It has been generally determined through testing, which type of fastener functions best in the more commonly used material types.

If a standard thread-forming or thread-cutting screw is recommended, some limited hole size recommendations may be found in the recommendation tables found in American National Standards Institute (ANSI) and

---

Material in this chapter has been both adapted and directly extracted from *Assembly Engineering*, August and September, 1982. By permission of the publisher © 1982 Hitchcock Publishing Co. All rights reserved.

FIG. 6.1   Many plastic products are molded with bosses for fastening
sites. Because bosses are made as small as possible to economize on
materials, they can burst. (From *Appliance Manufacturer*, © March
1979, Cahners Publishing Co., Inc., Boston, MA.)

Industrial Fastener Institute (IFI) standards, although the standards
may not cover the newest polymers. However, if a fastener especially
designed for plastics is to be used, then the hole-size recommendation
must be obtained from a manufacturer's brochures or from the manufacturer.
    Determination of proper hole size for critical applications often re-
quires prototype testing to determine if the proper clamp load, torque,
and pull-out strengths can be obtained. Some guidelines for estimating
these values are given later in this chapter.
    In the manufacturing process, pilot holes are formed by core pins
which usually have from a one-half to three-degree taper. This taper
permits removal of the pin after the molding process has been completed.
The degree of taper varies for different component designs and materials. When hole size recommendations do not consider taper, as is usually the case in laboratory reports, the recommended hole diameter must be
assumed to occur at one-half the hole depth.
    The pilot hole should be counterbored to a full or major screw diam-
eter and to a depth equal to one-fourth to one-half of the screw pitch.
This reduces the chance of chipping as the screw enters the hole, and
it permits easier starting of the screw in the thread-forming operation.
    Only rarely in the assembly process with self-threading screws are
holes formed by drilling into the plastic. In such cases the holes have
no taper. Drilling is commonly used in prototype construction and in
laboratory testing. However, the torque and tension values obtained
with drilled holes in certain materials such as polyphenylene oxide
(PPO) or polycarbonate (PC) will be slightly different from those obtained with cored holes.

Drilling through the fiber of a molded plastic part causes fiber fray-ing at the cut location. The fraying results in different test values caused by higher friction when the screw is inserted, also higher tor-que values. When analyzing test data obtained with drilled holes, these higher values should be considered by the laboratory in esti-mating probable cored-hole equivalents.

There is another situation which occurs specifically with long-fiber-filled materials. In this case, a different fiber orientation occurs on a molded part in locations where there are no holes, than around a mold-ed hole, where the fibers may tend to align themselves beneath and around the core pin.

Structural foams and reaction injection molded (RIM) materials have cellular structures when molded, and a hard exterior skin formation. Satisfactory testing cannot be done in these materials with drilled holes. Furthermore, in structural foam neither drilling nor use of thread-cutting screws should be considered for production as the cutting faces on the drill or screw will weaken the potential joint by cutting away the hard surface material.

6.1.2 Boss Diameter

Molded plastic components have internal stresses which are created in the molding process. These internal stresses are increased when a fastener is installed and tightened. (The amount of stress increase during assembly will of course vary with screw design.) The larger the boss diameter, the greater the area over which stresses can spread, thus reducing any tendency to crack the boss. With thread-forming or thread-rolling screws, as a general rule, boss outside diam-eter should be two to three times the hole diameter. Observing these parameters can also be helpful in reducing the detrimental effect of possible molding imperfections in the boss.

Some engineering-grade plastics also have critical boss thickness and wall thickness relationships. To avoid the appearance of "sink marks" on the exterior of the plastic, opposite the boss, the proper relation-ship must be maintained, particularly with polycarbonates. In an ex-posed location, such a mark could be cosmetically displeasing. Also, this relationship determines the maximum boss size.

When boss diameter is minimal, screws for plastics with extra-wide spacing of threads are preferred. The height of the boss, gusset, or other mounting site should also be designed to avoid any space between the part being fastened and the top of the boss, otherwise undue stress will occur in tightening.

6.1.3 Costs, Cosmetics, and Strength

Other boss-design considerations also will be involved and the designer must be prepared to make compromises. Material cost, molding cost, cosmetic appearance, and required joint strength have to be taken into account.

Large bosses use more material. Are they justifiable economically and weightwise? Use of large bosses also tends to increase molding cycle time and increases the risk of unsightly sink marks on the opposite face of the part.

6.1.4 Screw Engagement

The length of engagement refers to the threaded length that is actually engaged in the plastic. A well-designed fastening site should accommodate an engagement length two to three times the diameter of the screw. Screws can, of course, be installed to whatever depth is required. However, depths beyond three times the diameter of the screw usually result in high-drive torque, without any significant gain in assembly strength. In such cases, lubricizing of the screw may be considered; of course, absorption of the lubricizing element into the plastic must be considered, as well as its effect on the joint and on the appearance of the assembly.

6.1.5 Determining Screw Diameter

When the engagement length is three times the screw diameter, the stripping resistance of the joint will approach the torsional strength of the screw, and pull-out strength will approach the tensile breaking load of the screw. These values are used to estimate the screw size required for a particular application.

When using engagement lengths of three times the screw diameter, it is possible to compare the required, calculated part usage loads (estimated from standard engineering formulae) to the torsional strengths or tensile breaking loads of a range of the screws to be used. The latter values should be available from the manufacturer. Screw diameter choice can thus be greatly simplified.

6.1.6 Boss Design: General Remarks

Experts in the design of plastic parts suggest that to avoid sink marks it is advisable to remove the boss from the wall of the part and connect it to the wall with a thin rib, in thicknesses from 40 to about 80 percent of wall thickness, depending on material characteristics and cosmetic requirements. Consultation should be made with the resin supplier. Examples of boss designs are shown in Fig. 6.2.

By removing the boss location from the wall, use of a constant boss wall thickness is enabled without the temptation to place the pilot hole too close to the exterior of the wall, which may then crack.

Weld lines and gate locations can be the weak points in bosses where cracking occurs. To comprehend this statement it is necessary to understand how the plastic gets into the mold cavity. Molds for injection molding have to have entrances or channels through which the hot, liquid plastic is introduced. The channel leading into the mold is called

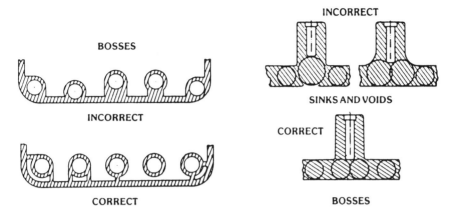

FIG. 6.2 Boss and wall thickness relationship can be critical with certain engineering-grade plastics. To avoid "sink marks" on the exterior
wall facing, resin producers tend to advocate boss designs similar to
those shown at the bottom. The designs shown in upper drawings may
be advocated by the fastener manufacturer for greater joint strength.
(From *Plastics Design Forum*, © July/August 1982 Industry Media, Inc.,
Denver, CO.)

the "runner" and the exact entrance from the runner into the mold is
called the "gate." Cooled plastic remaining in the runner after the
molding operation is complete will be snipped off or otherwise separated
from the plastic part at the gate which is generally of smaller diameter
than the runner.

    When the plastic flows into the mold cavity, it will enter through one
or more runners. Wherever the flowing plastic meets plastic coming
into the mold from other runners and gates, or when it flows around
a pin or other projection into the mold and meets itself, right there will
be a weld (or joining) line in the finished piece. This weld line will be
the weakest point in the material, even when the material contains fiber
filler, as no filaments will cross through the weld line. If the weld line
occurs down the length of the boss as it will usually do, then cracking
problems will exist. There are, however, gate design methods which
eliminate the problem. Also, if the gate is located too close to the pilot
hole, there can be cracking. For these reasons alone, designers should
be aware of molding methods and mold design.

6.2 ASSEMBLY CONSIDERATIONS: SELF-THREADING SCREWS

6.2.1 Drive and Fail Torque

One of the most common problems when using self-threading screws in
plastics is stripping during assembly. The relatively low strength of

	Drive	Fail	Seating torque range differential
Example 1	3 lb-in.	12 lb-in.	9 lb-in.
Example 2	5 lb-in.	20 lb-in.	15 lb-in.

FIG. 6.3   Theoretical comparison of 4:1 fail-drive ratio. Note that greater differential values generally lower the likelihood of failure. (From *Assembly Engineering*, © September 1982 Hitchcock Publishing Co., Inc., Wheaton, IL.)

many plastics can result in low torque values. In the application of screws in metal, achieving a satisfactory fail-drive ratio of 3:1 may be an important goal. However, in plastic assemblies, because of the lower torque conditions encountered, it is more important to achieve a satisfactory fail-drive differential (range of torque between drive and fail).

In the theoretical example shown in Fig. 6.3, example 1, the drive torque is 3 lb/in (.339 nm) with a fail torque of 12 lb/in (1.356 nm). This gives a differential of 9 lb/in (1.017 nm). After considering the statistical variance of the torque values and the setting accuracy of typical power-driving equipment, this differential does not allow much room for error. If the torque magnitude could be raised to 5 lb/in (.565 nm) drive and 20 lb/in (2.26 nm) fail, we would still have a 4:1 fail-drive ratio and the differential would be significantly improved. This would considerably reduce chances of assembly-line stripping.

As we indicated in Chap. 2, the screws which have been specially designed for use in plastics provide higher all-around levels of torque than do standard thread-forming screws. Therefore, they are generally preferred for high-production assembly. A typical increase in torque differential over standard thread-forming screws is indicated in Fig. 6.3.

With standard thread-cutting screws used in the recommended holes, drive and strip torques are relatively low. Control of driving equipment may be difficult. However, low torque is sometimes requested.

6.2.2   Torque Tension

The new, harder engineering grades of plastics are now being used in structural assemblies. With this in mind, fastener laboratories are studying torque-tension relationships. The special-design fasteners for plastics create more torque. As a result of the introduction of the new, harder materials and the new, specially designed screws, greater clamp load can be applied to plastic assemblies.

Plastics are viscous, elastic materials. The mechanical formulae used with metals for the computation of torque-tension relationships do not apply to these materials. As a result, fastener manufacturers now are testing screws used in plastics. One of their major concerns has been load relaxation of plastics when used in structural applications. Recent studies with these special-design screws in engineering grades indicate some similarity in load relaxation to that for metals. Of course, the engineering grades without fillers are far stiffer than commodity plastics without fillers and exhibit less "creep."

The axial, load-bearing capabilities of the fastening joint have also been of concern to investigators. As mentioned before, when using an actual engagement length of three times the screw diameter, the pull-out strength of the fastener approaches its' tensile breaking load.

6.3 ASSEMBLY EQUIPMENT: SELF-THREADING SCREWS

The subject of assembly equipment is covered in detail in Chap. 9.

Power-driving equipment used to drive screws in metal applications can generally be used to drive screws into plastics. However, excessive rpm should be avoided as the resulting high heat could cause considerable expansion of the plastic and consequent lower torque. At the present time, we are not in a position to define the term "excessive" used in this context properly, in terms of surface feet per minute. For lack of the data, however, we fall back on a figure of 1000 rpm as the borderline, over which we feel too much heat can be generated.

6.3.1 Power Screwdrivers

The simplest type of power driver is the "air-stall" type, which shuts off with excess air pressure when a predetermined torque level is obtained. This tool permits safe torque levels to be repeatedly attained; resulting tension is good. Air-operated and electric-powered equipment, when provided with an adjustable clutch-driving mechanism, is also very good for installing helically driven screws in plastic assemblies.

6.3.2 Requirements for Push-in Fasteners

When using the axially inserted "push-in"-type fastener (see Chap. 2, Fig. 2.10), any axial-pressure method is acceptable, whether hydraulic, pneumatic, or mechanical. In an increasing number of cases, special, highly automated equipment is being devised for multiple, high-production assembly. Of course, simple rivet-setting machines may be used. In our research we have noted that in most push-in threaded-fastener applications the axial force required to pull the fastener out of the hole exceeds the force originally required for installation.

Ultrasonic insertion of threaded push-in fasteners provides the opportunity to use the type of equipment previously reserved for use with the more expensive metal insert-machine screw system, for direct insertion of fasteners into plastic. It is especially recommended for the more fragile applications and should raise their ability to withstand installation shock.

Another way to avoid high insertion forces with push-in-type fasteners, and one which seems to have considerable promise, is through use of the high-power, portable, pneumatic equipment employed for rapid insertion of nails in building construction. The almost instantaneous insertion appears to generage less heat and have a reduced impact effect on the fastening site.

6.4 TROUBLESHOOTING: SELF-THREADING SCREWS

There are three major problem areas in the application of self-threading screws to plastics: (1) stripping during assembly, (2) cracking of the plastic material, (3) loosening of the joint because of material relaxation. The following suggestions will help eliminate these problem areas.

6.4.1 Stripping During Assembly

Solution. Increase the fail-drive differential by one or more of the following methods for increasing screw involvement.

1. Lengthen the screw to increase engagement length.
2. Reduce pilot-hole diameter or increase screw diameter. (Changing the core-pin sizes in the mold is much cheaper than changing the whole mold.)
3. Increase underhead friction.
    a. Use "torque robbers" (serrations or nibs under the head) if material damage can be kept to a minimum. With special under-the-head design, material damage can be reduced.
    b. If fastening a metal part to plastic, consider a change to a higher friction fastener plating. When the bottom of the screwhead comes into contact with metal, plating friction can be effective. It will, however, have little effect when coming into direct contact with most plastics. Exceptions can be filled materials.
4. Change screw design.
    a. Use a double-lead thread.
    b. If thread cutters are being used, consider a thread-forming screw.
    c. If standard thread formers are being used, consider changing to screws specifically designed for use in plastics.

6.4.2 Cracking of the Plastic Material

Solution. Reduce (or spread out) induced stress at the fastening site by one or more of the following methods.

1. Reduce drive torque.
    a. Use screws designed especially for plastics that have:
        thread angles less than 60 degrees and/or
        extremely wide thread spacing and/or
        trilobular-shape design.
    b. Enlarge pilot hole (only if no detrimental effect on torque or pull-out performance occurs).
2. *Enlarge boss area to better distribute stress. (With screws, the minimum boss OD should be 2.5 times the pilot-hole diameter, or greater, except in very ductile materials where 2.0 may be sufficient.)*
3. Use a screw with a larger surface under the head or add a washer.

6.4.3 Loosening of the Joint

Material relaxation is inevitable in plastic materials. Either it must be minimized or a means must be devised to keep the fastener in place, despite its presence.
Solution

1. Use the specially designed screws for plastics that feature trilobular design. The cold flow or creep of the plastic will cause it to flow into the relief areas between the lobulations. This will help reduce stress and add locking power to the assembly, and/or
2. Minimize thickness of plastic being compressed (the part being fastened). In reducing the compressed cross-section, creep magnitude is reduced, and/or
3. Spread out the applied, compressive load on the plastic from tightening the fastener by using a fastener with a larger under-head bearing surface. This can be done by using a larger integrated washer on the head or by using a flat washer. The result is wider distribution of the load.
4. If the material can be changed to that with the next higher filler content (perhaps from $7\frac{1}{2}$ to 15% glass reinforcement), added stiffness and less creep will result. If existing tooling is to be used, then care must be taken to match the material shrinkage.
5. Max Augenthaler of Gobin Daude, S.A., told us about a variation in this problem, as when fastening a soft plastic part to a metal nut member, and recommended the following solution. Under the circumstances described, the tendency is for the plastic under the head of the screw to compress and flow outboard. Such movement can be reduced and some of the plastic trapped

under head by providing the bottom edge of the screwhead with a concentric tooth or ring encircling the outboard underside of the head.

6.4.4 General Considerations

There is one more area for consideration when troubleshooting; it is so obvious perhaps we should have mentioned it first.

Does the manufactured part meet the materials specifications as to composition and design at the fastening site? In the next chapter we give examples of what can happen when they do not.

6.5 TAPPING IN PLASTICS

The primary problem encountered in tapping in plastics (indeed, with all machining of plastics, especially drilling) is the frictional creation of detrimental heat from the elastic recovery and rubbing action of most of these materials. Too much heat will be destructive to the materials surrounding the threaded joint. The use of ground thread, polished flute taps manufactured from high-speed (HSS) alloy steels (AISI M-1, M-7, M10 or the more expensive M-3; ISO S2, S-3) are suggested.

6.5.1 Tapping in Through Holes

Hard plastics such as polycarbonates and plastics which have been loaded with minerals, etc., tend to produce a powdery chip. Under this condition for a volume-tapping operation where heat buildup from friction is a consideration, recommendation from knowledgeable individuals calls for use of special, multiple flute taps with narrow lands. The purpose of the narrow lands is to help keep down the heat buildup. Generally with no. 6 diameter and larger, five flutes are recommended.

Everyone knows that in machining in metals, tools must be kept sharp. This is equally important when tapping plastics. Consider the effect of adding glass filler content. Even without fillers, the wear will be considerable. Before use, taps should be inspected for freedom from burrs.

For working in plastics, taps should be provided with a surface treatment to reduce abrasive wear. In the past, nitriding, hard chrome, or a combination of nitriding and hard chrome have been utilized. Just coming on the market are the new tooling coatings, such as titanium nitrides. In our own experience, these coatings provide dramatic tool life increases in the manufacture of metal screws. They certainly should be considered for taps used on a volume basis in plastics.

Through-hole tapping in ductile materials that tend to produce "stringy" chips will best be accomplished with spiral *pointed* taps (Fig. 6.4) which have been surface-treated.

FIG. 6.4 Spiral-pointed plug tap for use with "stringy" chip-forming materials in through holes. The angular grind on the point drives the sheared chips forward and out of the hole. (Courtesy of Heli-Coil Products, Div. of Mite Corporation, Danbury, CT.)

6.5.2 Tapping in Blind Holes

For ductile materials producing stringy chips, spiral *fluted* taps (Fig. 6.5) are recommended when tapping in blind holes. If not available, the criteria for tap selection should be based on obtaining the maximum in flute area to make room for the sinewy type chips. If necessary the flutes can be further ground. In deep-hole tapping, taps may have to be backed out to prevent excessive chip accumulation and use of tapping lubricants or air blasts may be helpful. With the hard plastics any style of tap is satisfactory.

6.5.3 General Remarks on Tapping in Plastics

Tapping fluids are generally (not always) used to keep down heat. The fluids should not contain chemicals that will be destructive to the plastic.

In considering thread types, notch sensitivity is a factor to be taken into account with many thermoplastics, and for this reason the taps should have truncated threads such as are provided by the American Standard unified thread form.

Another factor to be considered is the tendency of ductile plastics to cold flow after the tapping operation. Fortunately, in order to rectify the situation taps can be obtained in standard oversize dimensions, such as H-3 for the smaller sizes or H-5 for the larger sizes.

Rake and clearance angles may differ considerably between various plastics. It is recommended that the resin producers' design manuals be consulted.

FIG. 6.5 Spiral fluted bottoming tap used in pulling stringy chips from ductile materials out of bottoming holes. (Courtesy of Heli-Coil Products, Div. of Mite Corporation, Danbury, CT.)

Peripheral tapping speeds for HSS taps as compiled by the technical staff of the U.S. Department of Defense are listed as follows (1):

Unfilled materials, 50 ft/min (15 m/min)
Filled materials, 25 ft/min (8 m/min)

A short and excellent general source on drilling, tapping, and all kinds of machining in plastics is The Polymer Corporation's *Plastics Fabrication Manual*; see Appendix 1 for address. They have offices throughout the United States and in nine countries. Specific machining information on individual plastics is also available from the product manuals provided by the leading resin producers. A first-class and considerably detailed source of information is *Plastics Engineering Handbook* of the Society of the Plastics Industry, Inc. a lot of unnecessary headaches will be avoided by reading these manuals before beginning a tapping operation.

REFERENCE

1. *Machining Data Handbook*, vol. 1, 3rd ed., Machinability Data Center, Cincinnati, 1980, pp. 4–41.

BIBLIOGRAPHY

Gomes, K. J. *Assembly Engineering* 25:26-29 (September 1982).

Farrah, M. *Plastics Design Forum* 7:59-68 (July/August 1982).

Polymer Corporation, The. *Plastics Fabrication Manual*, 1982.

Reynolds, R. L. *Assembly Engineering* 14:26-30 (September 1971).

Society of Plastics Industry. *Plastics Engineering Handbook*, Van Nostrand Reinhold, New York.

Wagner, D. *Assembly Engineering* 23:26–28 (November 1980).

7
TYPICAL APPLICATIONS

The following applications were analyzed by the product engineering laboratory of a major U.S. fastener manufacturer. In each case, the application was requested by a manufacturer involved with plastic assemblies or by the company's design staff. Some of these applications illustrate the bind a manufacturer can get into by following preconceived, and not necessarily correct, assumptions.

7.1 MACHINE SCREWS VERSUS SPECIAL-DESIGN SCREWS FOR PLASTICS

The material used in this electrical junction box is an unspecified, somewhat ductile resin, highly filled with glass fiber. The application is somewhat unusual for plastics because the manufacturer had been using machine screws, fabricated in plant and inserted directly into pretapped holes in the plastic. The performance results using machine screws were not particularly favorable, and the manufacturer asked for assistance.

Performance standards included a pull-out requirement of 50 lb per screw (2 screws per box) to withstand the weight of a heavy light fixture. Failure torque requirements were set at 20 lb/in (2.26 nm). These values are low, as befitting use of a machine screw. Figure 7.1 shows that the design and construction of bosses, fastening sites, and the box itself were excellent and amenable to the use of thread-forming screws.

Material in this chapter has been both adapted and directly extracted from *Assembly Engineering*, August and September, 1982. By permission of the publisher © 1982 Hitchcock Publishing Co. All rights reserved.

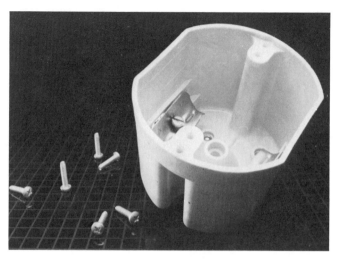

FIG. 7.1   Replacing machine screws with special-design thread-rolling screws for plastic in this electrical junction box not only provided joints of far greater structural strength, but also substantially reduced in-place fastening costs.

Application Study 1

The fastener manufacturer's laboratory was asked to test their company's newly designed special thread-rolling screw for plastics and provide test data for the electrical junction box manufacturer. Laboratory test comparisons (Fig. 7.2) were determined using a graphic amplifier-recorder. Torque was sensed by a spindle-mounted transducer. The laboratory equipment used is illustrated in Chap. 8, Fig. 8.3.

Screws were statically driven using a hand-held wrench connected to the transducer. Plain, flat steel washers were used under the screws to act as spacers, to accept clamp load, and to simulate the lighting fixture being attached. The use of thread-forming screws has eliminated the tapping operation which was originally being done in the four visible holes at the top of the box. Savings from this operation alone were estimated in excess of 60 cents per unit. Pull-out tests were not performed in the laboratory comparisons, since with a minimum of 1/2 in. length of screw engagement for the newly designed screw, pull-out strength should approach the screw strength of 1000 lb or more.

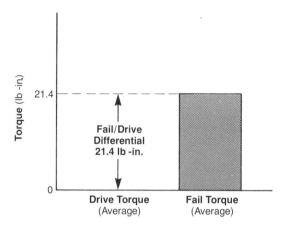

#8-32 x ⅝″ Slotted Round Head Machine Screws, Zinc Plated, into Fiberglass Filled Resin, .137″ Diameter Hole

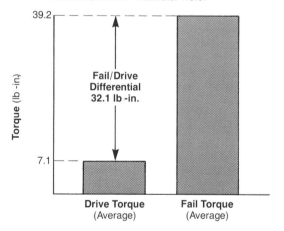

#8-11 x ⅝″ Hex Washer Head Special Screws for Plastic, Cadmium Finish into Fiberglass Filled Resin .137″ Diameter Hole.

FIG. 7.2 Graphically portrayed comparison of the 50% greater fail/drive differential obtained by the use of special-design thread-rolling screws for plastic. Use of such screws not only provides improved installation but saves a substantial amount of expense, as indicated in the text.

7.2 SCREW SELECTION AFFECTS AN AUTOMOTIVE ASSEMBLY LINE

This valve housing (Fig. 7.3), used with automobile bucket seats, is manufactured from 33% glass-filled Nylon. A 1/8-inch thick steel channel with clearance holes is fastened to the housing with four screws. The application is another example of how screw selection can affect fastening results—this time on the assembly line.

The manufacturer had been assembling the product using a proprietary thread-rolling screw which appeared to give excessively high drive torque. Design requirements prohibited the use of a smaller size of the same screw.

Application Study 2

Test results, shown graphically in Fig. 7.4, were obtained with the same procedure outlined for application study 1. Torque was sensed by a spindle-mounted transducer and recorded on the laboratory's graphic amplifier-recorder. Screws were statically driven using a hand-held wrench connected to the transducer. Plain, flat steel washers were used under the screws to act as spacers, to accept clamp load, and

FIG. 7.3 The use of special, trilobular-shaped, thread-rolling screws as replacements for other proprietary thread-rolling screws in a valve housing eliminated the problem of excessive drive torque without sacrifice of strength in screw or assembly.

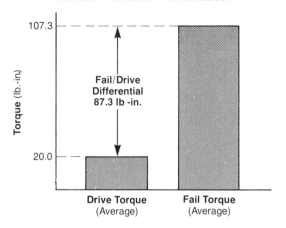

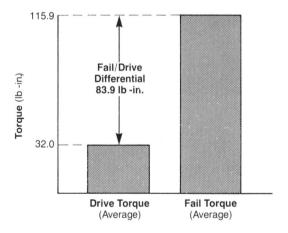

FIG. 7.4 Graphic comparison of the 37.5% reduction in drive torque obtained in switching from one proprietary screw to another, without sacrifice of joint strength. (From *Assembly Engineering,* September 1982, © Hitchcock Publishing Co., Wheaton, IL.)

to simulate the lighting fixture being attached. The laboratory equipment used is illustrated in the next chapter in Fig. 8.3.

The laboratory, which tested both the existing proprietary screw and their company's own special design, concluded that the use of their special, trilobular-shaped fastener shape would eliminate the high-drive torque without sacrificing the required strength in the screw or the assembly. As can be seen in Fig. 7.4, there was a 37.5% reduction in drive torque, offering the opportunity for substantially improved performance on a rapidly moving automotive assembly line.

7.3   FAILURE TO CHECK OUT THE FASTENING SITE WITH A LABORATORY

Application study 3 covers an acrylic instrument panel used on an appliance housing assembly. The problem in this case was caused by failure to contact a fastener laboratory at the product design stage. A basic difficulty faced by designers has been the lack of available information on the use of threaded fasteners in plastics.

The manufacturer of the appliance had been using a type "F" thread-cutting screw since standard thread-forming fasteners are not effec-

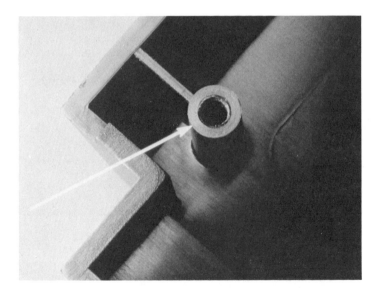

FIG. 7.5   This hard, rigid acrylic instrument panel had .124–.120 in. diameter taper-cored holes in .255 in. diameter bosses. This provides a boss OD-to-hole ratio which is less than the 3:1 recommended and could lead to boss cracking unless otherwise compensated for. (From *Assembly Engineering*, September 1982, © Hitchcock Publishing Co., Wheaton, IL.)

tive in hard, rigid materials such as acrylic. The no. 6-32 × 3/8 in screws were being driven into .124—.120 in diameter, taper-cored holes in a .255 in. diameter boss. This provided a boss OD-to-hole ratio of about 2:1 which is less than the 3:1 recommended. (See Fig. 7.5.)

At a 3:1 ratio, the boss would have been large enough to disperse induced stresses. However, the manufacturer stated that the fastening site could not be changed—probably due to an existing inventory of already manufactured parts.

Because of difficulties associated with the use of the thread cutters previously described, a fastener manufacturer's laboratory was asked to test their special new thread-rolling screw design. Test results follow:

Application Study 3

In the following tests, No. 6-13 special-design thread-rolling screws for plastic, with a phosphate and oil finish, were used with .255 in. diameter acrylic bosses. (Torque values are in lb-in.)

	Hole no.	Top hole dia.	Bottom hole dia.	Depth of hole	Drive torque	Prevailing off torque 1st	5th	Fail torque
Blind	1	.123	.119	9/16	2.6	3.0	.7	13.6 Hole
holes	2	.124	.120	9/16	2.4	2.2	.9	12.1 stripped
Thru	3	.124	.122	5/16	2.6	2.5	.7	12.2 Hole
holes	4	.124	.122	5/16	1.8	2.0	1.0	12.9 stripped
				Aver.	2.35			12.7
				Sigma	.38			.7

Holes 1 & 2 - 3/8 in engagement
Holes 3 & 4 - 5/16 in engagement

Fail/drive ratio 5:4   Differential 10.35   No boss cracking on drive or fail

In the tests, the screws were driven at 300—400 rpm while maximum torque was recorded. The rotation was then reversed, recording first prevailing torque. Screws were cycled in and out four additional times, recording the fifth prevailing torque. The screws were then driven to failure. Prevailing (off) torque, a fastening's resistance to free turning, is an indication of its resistance to vibrational loosening.

Plain flat steel washers were used under the screw heads to accept the clamp load, simulate actual application, and achieve proper engagement length. Torque was sensed by a spindle-mounted, dynamic transducer and was monitored on the digital display of a peak meter. Screws were driven using a pneumatic pistol-grip driver.

7.4   EFFECT OF DIFFERENT MOLDING TECHNIQUES

As we have mentioned before, variations in plastic materials may occur because of differences in molding and manufacturing techniques from one supplier to another—even on the same resin—and even from batch to batch. Such a situation may have occurred in the following application study. In this case the fastener laboratory was asked by the resin producer, to test the use of a special-design screw for plastics in two different types of well-known videocassettes which had been manufactured from acrylonitrile-butadiene-styrene (ABS) plastic (Fig. 7.6).

Application Study 4

The cassettes were being made and assembled in Europe and shipped to the United States. Number 4 diameter, standard-type BT thread-cutting screws were being used in the assembly.

At the laboratory, tests were performed on some of the cassettes using the existing fasteners, as received assembled in the cassettes. These tests were conducted in three ways:

FIG. 7.6   The fastening characteristics in two types of similar cassette housings, both manufactured from the same type of plastic were found to be significantly different. Since injection molding techniques vary from molder to molder and since the techniques have a very significant effect on a plastic's performance, it is wise to stick with a single molding source.

Cassette type A: No. 4 × 1/2 in. type BT, Phillips pan-head thread-cutting screws, as received assembled in the cassette—ready for testing.

Test 1:

Breakaway torque	Prevailing off torque	Fail torque
1.2	.72	12.0
1.2	.60	11.3
1.4	.72	11.0
1.4	1.00	11.4
Avg. 1.3	.76	11.4

Test 2: Screws as received assembled in type A cassettes were seated, broken away, and driven to failure.

	Seating torque	Breakaway torque	1st Prevailing off torque	Fail torque
	8	4.8	.72	11.4
	8	3.3	.80	11.4
	8	5.0	.72	11.4
Avg.		4.4	.75	11.4

Test 3: Screws as received assembled were cycled in and out for five cycles in type A cassettes, then driven to failure.

	Seating torque	Breakaway torque	Prevailing off torque 1st	Prevailing off torque 5th	Fail torque after 5th cycle
	8	2.6	1.4	1.10	5.2
	8	2.8	1.4	.60	4.8
	8	2.9	1.2	.72	4.1
	8	3.0	1.4	.72	4.8
	8	2.6	1.2	.48	5.9
	8	2.6	1.3	.72	6.0
Avg.		2.8	1.3	.72	5.13

Cassette type B: No. 4 × 1/2 in. Phillips pan-head, special-design thread-rolling screws for plastic, driven into .096 in. previously drilled holes.

FIG. 7.7 Test results (torque values in lb-in.).

Test 4:

Drive torque	Seating torque	Breakaway torque	Prevailing off torque 1st	Prevailing off torque 5th	Fail torque after 5th cycle
1.9			1.7	1.1	12.0
1.8			1.7	1.2	13.4
1.9			1.6	1.2	12.7
3.4			3.4	1.8	11.6
3.0			2.6	2.3	12.0
3.1			3.0	1.7	12.0
2.4			2.4	1.9	14.6
1.9			1.6	2.2	14.4
Avg. 2.4			2.2	1.7	12.8
3.0	8	5.0	1.6		11.7
3.1	8	5.3	1.4		14.7
2.8	8	4.8	2.2		11.7
3.0	8	4.4	1.7		14.7
Avg. 2.9		4.9	1.7		13.2

Test 5: Screws as received assembled were cycled in and out five times then driven to failure in type B cassettes.

			1.7	.5	11.4
			2.3	1.0	9.1
		Avg. 2.0			

Test 6: No. 4 × 1/2 in. Phillips pan-head, special-design thread-rolling screws for plastics in type B cassettes. The no. 4 thread-cutting screws were removed and existing holes were redrilled to .096 in. diameter to accept the slightly larger special-design screws being tested. The full potential holding power of the special-design screws for plastics would have been better served by insertion in unused pilot holes, but unassembled cassettes were not available.

			3.1	3.1	13.7
			3.8	2.4	14.4
Avg. 3.4			3.5	2.8	14.1

FIG. 7.7 (Continued)

1. On several of the cassettes, back off and prevailing off torques were recorded on existing screws, and then driven to failure.
2. On other cassettes, the thread cutters were cycled in and out five times, recording prevailing torques without seating the screws. These screws were then driven to failure.
3. Also, some no. 4 thread cutters were seated at 8 lb-in., which resulted, it was noted, in significantly higher breakaway torque.

Following these tests, the laboratory used their company's own special-design screws for plastic, which happened to be spaced-thread, single-lead screws with a 60° thread angle.

Results are shown in Fig. 7.7 for both types of cassettes, and it will be noted that the results are quite different, although both types of cassettes are made from ABS plastic.

This is a good example of why knowledgeable designers try to work with a single source for their parts. It also demonstrates how different processing techniques or conditions may affect parts.

7.5 OBTAINING A CONSISTENT MAGNITUDE OF TORQUE WITH SMALL SCREWS

Here is a classic problem that plastic parts manufacturers sometimes run into with very small screws. With very small screws, the magnitude of torque can be so low that many driving systems cannot repetitiously supply torque with enough consistency.

Application Study 5

The product manufacturer, in this case a well-known computer company, was experiencing stripping problems using a no. 2 diameter, specially designed thread-rolling screw for plastics. The screw held together the lips of both halves of a magnetic tape drive (Fig. 7.8) made from a leading polycarbonate.

The problem presented to the fastener manufacturer's laboratory was twofold; first, to salvage stripped disc drives still on the production floor, and second, to provide a final answer for continuing production. The answers to both problems will be found among the suggestions in Chap. 6, Sec. 6.4.1.

The repair problem was a simple one; the laboratory sent the manufacturer a supply of the next larger diameter screw to be inserted in existing holes. Their suggestion for a long-term solution also permitted use of the original pilot holes, and this was to use a no. 2, double-lead thread-rolling screw for plastics. Figure 7.9 illustrates a 23.7% resulting increase in differential that was achieved without any need to change pilot-hole diameter. This is an example of our contention that there is no one type of fastener that will solve all fastening problems in plastic materials.

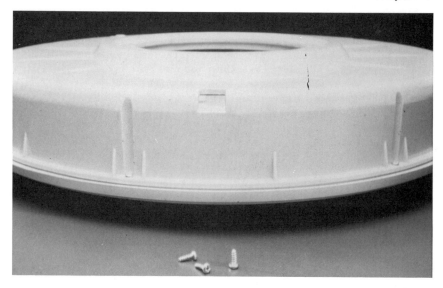

FIG. 7.8 The magnitude of torque on small screws used in plastics
often is so low that stripping becomes a problem.

Standard 60° thread-forming screws vs.
special-design, trilobular-shaped thread, double-lead
thread-rolling screws in polycarbonate

	Standard 60° screw	Special-design double-lead screw for plastic
Average fail torque, lb-in.	4.8	5.8
Deduct:		
Average drive torque, lb-in.	1.0	1.1
Differential, lb-in.	3.8	4.7
Fail/drive ratio	4.8:1	5.8:1

23.7% Increase in differential

FIG. 7.9 Comparison of thread-forming screws.

FIG. 7.10 Because of internal hydraulic pressure in this structural installation, length of engagement used on the fasteners is greater than normal.

7.6 A PROBLEM JUST THE OPPOSITE OF THE USUAL

This application offers a real structural problem of unusual fastening difficulty. It had a solution which we do not very often recommend; in fact, we do not include it under the suggestions in the previous chapter. The application, an extremely high-quality water softener (Fig. 7.10), probably could not have even been successfully manufactured from plastics before the development of the stronger engineering grades. Design requires offsetting a significant amount of internal hydraulic pressure.

The water softener was to be manufactured from GFN-3, Noryl with 30% glass fiber. As can be seen from the illustration the screws are long; and there are a great many of them, as required to withstand an hydraulic pressure of 150 psi. This equates to an axial load of 445 lb on each screw in the main assembly. Because of the structural requirements, length of engagement is higher than usual—3-to-4 times screw diameter—rather than the normal 2-to-3 times.

FIG. 7.11 Testing procedures in the laboratory can utilize transducers in recording data. Sophistication of currently available laboratory equipment as discussed in Chap. 8 can add immensely to the ability to test fastened assemblies under simulated field conditions.

No. 10 (.208 in.)—14 × .875 in. indented hex washer-head, 60° single-lead special-design thread-rolling screws for plastic, cadmium finish in GFN—3 Noryl with 30% glass fiber.

Maximum drive torque	Prevailing off torque	Fail torque	Type of failure
.750 in. engagement—nonlubricized (actual product condition)			
25.0	24.0	89.34	Hole stripped
26.0	25.0	102.5	Screw broke
23.75	23.75	88.75	Screw broke
26.25	26.25	100.0	Screw broke
27.5	26.25	103.13	Screw broke
26.88	25.0	91.88	Hole stripped
23.75	17.5	103.75	Screw broke
26.25	27.5	103.75	Screw broke
Avg. 26.67		97.89	
Fail—drive ratio: 3.67:1		Fail—drive differential 71.22	
.750 in engagement—lubricized			
9.38	9.25	76.25	Hole stripped
11.4	11.4	85.0	Hole stripped
11.2	11.0	85.0	Hole stripped
11.2	11.2	74.38	Hole stripped
11.4	10.7	84.38	Hole stripped
10.0	10.0	77.5	Hole stripped
12.1	11.9	71.88	Hole stripped
14.0	14.9	87.5	Hole stripped
Avg. 11.34		80.24	
Fail—drive ratio: 7.08:1		Fail—drive differential 68.9	

FIG. 7.12   Torque values (lb-in.)

Application Study 6

After trying out a special-design screw, the manufacturer complained of high drive effort. This is just the opposite of the usual situation with plastics where, as illustrated in the previous application example, the problem is often one of low drive torque.

The laboratory set up their usual testing procedure. Torque was sensed from a spindle-mounted transducer (Fig. 7.11), monitored on

a peak meter, and graphically recorded on a dual-channel, brush am-
plifier test unit (see Fig. 8.3). The screws were tested at both a .750
and .500 in. length of engagement and then it was decided to see the
effects of adding a dry wax coating (Fig. 7.12).

Waxing, although a frequent resort of fastener manufacturers with
metal installations, is generally to be avoided with plastics and should
never be done without consultation with the resin supplier. In this

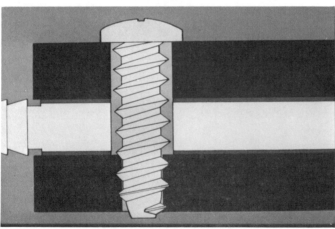

FIG. 7.13 Carburetor manufactured from Ryton (R4) plastic.

case, however, the material has chemical resistance to the lubricating element used. We do not recommend using phosphate and oil finishes with plastics. Lubricants used on fasteners may induce cracking in plastics if they are not compatible.

The addition of the wax in this application had another effect which we will discuss in Chap. 9.

7.7 THE IMPORTANCE OF CORRECT BOSS DESIGN

Like the material used in the preceding study, Ryton R4 polyphenylene sulfide has high chemical resistance. It also has high thermal stability and flammability resistance. Otherwise, it would never have been selected to be the material for this small, plastic carburetor (Fig. 7.13). It combines a hard, rigid structure with both high tensile and flexural strength; and in fact, although a thermoplastic, displays many of the thread-rolling characteristics of thermosets. The carburetor is intended for small, gasoline-powered motors such as are found in hand-held garden appliances and chain saws.

Design of the product is simple and neat as befits the prize-winning plastic carburetor designs of the manufacturer. In this example, the internal assembly member has a tight fit. If it had not, there could have been a cracking problem if the screws are overtightened, since design of the outboard assembly part leaves a gap between the fastening members. In this case, an internal part fills the gap tightly.

It is important that bosses, gussets, and all mounting sites which accept screw threads in plastic be designed to avoid any space between the part being fastened and the top of the boss in order to prevent cracks from occurring during tightening. This does not imply that boss design, especially in notch-sensitive materials, should not be provided with a counterbore (see Fig. 6.1). There should not be a gap, however, between fastened part and boss top.

7.8 WHEN SCREW SELECTION IS NOT THOROUGHLY TESTED

This example may be construed to illustrate what can happen if fasteners are selected without testing by a reliable laboratory. We are not 100% sure this is what happened because honest mistakes do occur in recommending or supplying screws. But in any case, in order to avoid embarrassment we are not going to identify the product very closely, other than to say that the product was manufactured from unreinforced polycarbonate and the parts manufacturer found himself with a plentiful supply of cored parts and a supply of spaced thread-forming screws that did not do the job.

Application Study 7

Pilot holes had been cored to approximately .157 in. The no. 8—18 ×
1/2 in. long screws being used were measured at .162—.160 in. diameter by the laboratory which the manufacturer called on to get him out
of the situation.

This diameter does not provide very much depth of thread (perhaps
25—30%) and is an indication of poor holding power. It is assumed that
such a large hole was found necessary to reduce stress, boss breakage, and high drive torque.

The newly contacted laboratory looked over the original screws, put
them aside, and tested their own company's no. 10—9 × .750 in. special-design thread-rolling screw for plastics which is highly suitable
for the existing .157 in. cored pilot holes. Indeed, this hole size is
right in the middle of its recommended diameter range for use in plastics.

The screw, which has fewer threads per inch than the original, size-
for-size, has only 9 threads per inch in the no. 10 diameter versus
18 in the no. 8 diameter thread-forming screw; however, the threads
are higher in the special-design screw and in this installation provided
approximately 60% depth of thread in the plastic.

No. 10—9 × 3/4 in. hex washer-head, 45° single-lead special-design
thread-rolling screw for plastic, cadmium finish, statically driven into
holes in unreinforced polycarbonate, as follows:

Hole	Top	Bottom	Depth
1	.158 in.	.157 in.	1/2 in.
2	.157	.156	1/2
3	.157	.156	7/16
4	.157	.156	7/16

Engage-ment	Hole	Drive torque	Prevailing off torque 1st	5th	Fail torque	Type failure	Fail-drive ratio	Differ-ential
3/8 in.	1	11.1	11.5	6.4	56.1	Hole	5.1:1	45
3/8	2	11.4	11.7	7.5	49.2	stripped	4.3:1	38
5/16	3	—	—	3.9	42.8	in all	—	—
5/16	4	6.8	6.3	4.5	32.4	cases	4.8:1	26

FIG. 7.14   Test results: torque values (lb-in.).

The screw design is intended to induce minimal stress in highly stressed plastics, such as polycarbonate. Indeed, the narrow profile and the extremely wide spacing do reduce lateral stress.

In the test the special screws for plastics were statically hand driven. Maximum drive (thread-forming) torque was recorded. Rotation was reversed, recording first prevailing off-torque. The screws were then cycled in and out four additional times and fifth prevailing off torque was recorded. Then the screws were driven to failure. Plan flat steel washers were used under the head to accept clamp load and achieve approximately 7/16 in. of engagement and to simulate the component being attached. As usual, torque was sensed from a spindle-mounted transducer and graphically recorded on a brush recorder. Test results are shown in Fig. 7.14.

7.9 A CASE STUDY IN STRUCTURAL FOAM AND SOME FAMOUS LAST WORDS

Here we have an example of a successful product design in structural foam manufactured from a highly regarded engineering-grade plastic. The parts manufacturer was using the screws that the resin producer recommended. However, after a considerable, highly successful production run, the manufacturer's composure was somewhat shattered when the assembly line started to have a great many boss cracks.

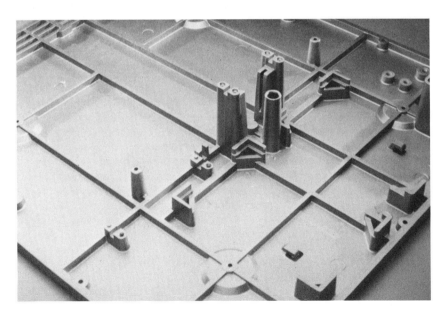

FIG. 7.15   Base plate.

Application Study 8

The product manufacturer, a well-known maker of cash registers, was baffled and contacted the molder. The molder stated that the pilot holes were too small and the screws were inducing failure. However, the register manufacturer would not "buy that," especially after examining holes and screws and finding both satisfactory. The manufacturer sent two of the base plates to the fastener laboratory (Fig. 7.15). One of the plates had cracked bosses.

In the laboratory, examination of the cored pilot holes did reveal a considerable variation in size, generally tending in the direction of unnecessarily large holes. Although this would account for variances in drive torque it was not a condition to cause boss cracking.

Careful inspection of the plate with cracked bosses was then made, and on the large boss which was cracked the inner structure could be seen, not cellular, but solid. Further inspection of the surface of the base brought to light a considerable amount of surface crazing in locations far from the fastening sites. At that point, the laboratory decided to bring in the resin producer and showed the base plates to the producer's structural foam engineers.

Examination by these engineers resulted in the conclusion that improper molding techniques had resulted in areas with resin-rich material and little cellular structure. Insufficient amounts of the inert gas or improper distribution of the gas can result in this situation.

The study is another example of a way in which material composition can affect fastening.

7.10 AUTOMATIC ASSEMBLY DEMANDS CLOSE ATTENTION TO FASTENER SPECIFICATIONS

The introduction of automatic or robotic screw-driving equipment into the assembly scene requires even greater attention to inspection of fasteners and parts. We will go into more detail on this subject in the following chapters, but now we are going to provide an example of one assembly problem that can arise.

In this case study, the product consisted of two kinds of auxiliary lights of the type that are often found on the highway, marking road equipment. The lenses were made from transparent-colored polycarbonate and were provided with through holes. They were fastened to either one of two different housings fabricated from ABS plastic. One housing had four bosses, counterbored 1/4 in. deep with .152–.145 in. taper-cored pilot holes. The other housing's pilot holes measured .149–.146 in. and were counterbored 1/32 in. deep.

Counterboring of the boss is the recommended procedure with plastics.

FIG. 7.16   The use of automatic screw-driving equipment demands
that the fastener provide consistent performance.   (Courtesy of Black
& Webster, Inc., Waltham, MA.)

Application Study 9

The manufacturer complained of inconsistent driving with the standard
type AB thread-forming screws being used with automatic driving equipment. In Fig. 7.16, a highly regarded automatic bench-type driver of
different manufacture is shown in the process of inserting screws. The
laboratory was asked to investigate whether their company's special-
design trilobular thread-rolling screws for plastic could be used to
alleviate the inconsistency in driving.

Tests for maximum drive, prevailing off torque and fail torque were
conducted by driving the special-design screws at 400 rpm with a
pneumatic screwdriver in the manner described in our previous case
studies. Test results follow.

Because of the difference in thread engagement in the two boss designs, installation torques would vary between them if the same screw
was used in both applications. Therefore, the laboratory recommended
using the lengths shown in the following table, as this procedure would
tend to equalize the required drive torques.

Automatic driving units require screws of consistent size, quality, and
performance to obtain efficient operation.

Maximum drive	Prevailing off torque	Fail torque	Type of failure

No. 8—16 × 3/4 in Phillips pan-head, cadmium-plated special-design, 60°, single-lead thread-rolling screw for plastic into .152—.145 in. taper-cored hole in ABS plastic; thread engagement 1/4—5/16 in, recess cammed out.

4. 7	5. 0	37. 6
5. 5	5. 8	29. 5
6. 7	6. 5	32. 5
6. 8	7. 2	32. 3
5. 2	6. 0	29. 2
5. 6	5. 5	32. 4
5. 5	6. 0	35. 4
5. 8	6. 0	31. 8

Average drive torque: 8 lb-in., average fail torque: 32.6 lb-in., Average fail/drive ratio: 4.07:1, average differential: 24.6 lb-in.

No. 8—16 × 7/16 in. Phillips pan-head, cadmium-plated special-design, 60°, single-lead thread-rolling screw for plastic into .149—.146 in. taper-cored hole in ABS plastic; thread engagement 1/4 in., recess cammed out.

7. 1	9. 0	37. 2
7. 6	9. 5	31. 6
8. 2	10. 1	31. 0
8. 9	10. 0	30. 2
7. 7	9. 5	29. 6
7. 8	8. 0	29. 3
7. 1	7. 8	34. 2
7. 3	9. 5	31. 4

Average drive torque: 7.7 lb-in., average fail torque: 31.8 lb-in., Average fail/drive ratio: 4.13:1, average differential: 24.1 lb-in.

BIBLIOGRAPHY

Gomes, K. J. *Assembly Engineering* 25:28— 29 (September 1982).

8

THE FASTENER LABORATORY

In this chapter we discuss the needs for and the functions of a fastener laboratory, the kinds of information a reliable laboratory should be expected to provide, and the kind of equipment one laboratory uses, and we will give an example of a continuing program which took place over several years between such a laboratory and an important organization in the field of electronic communications.

8.1 THE NEED FOR RELIABLE LABORATORIES IN THE FASTENING INDUSTRY

By this time, it should be clear to the reader that fastening in plastics is not a "cut and dried" proposition and that at this stage in the development of reliable techniques for fastening in the materials which have been developed so far, there is a dearth of readily available information. This is an important reason from the users point of view why fastener labs are handy things to have around, and as new materials are developed this will be a reason for their continued desirability.

However, good fastening laboratories are expensive to maintain and there may be a tendency on the part of some fastener manufacturers' managements to put less emphasis on them, indeed to eliminate them. Therefore, we would like to briefly remind management of what has happened over the past few years in the fastener industry in the United States and in other industrially developed countries.

For quite a number of years the historical markets for the North American fastener industry have been increasingly invaded by manufacturers from other countries. In this market seizure, overseas suppliers have been aided by their governments through direct or indirect subsidies. In addition, they often have had the advantage of lower labor costs, as well as newer machinery and better plant layouts. Distributors of fasteners have naturally been eager to cooperate because of lower prices, and in many cases, better service.

In the United States, the majority of standard fasteners sold come from overseas. The specialties market is being invaded as well. Not only has the U.S. fastener industry been backed into a corner of having to concentrate on purely special products, but even this corner is being attacked. In the future, even distributors may be increasingly so attacked by international exporting firms who are willing to go to the expense of supplying technical advice on a higher level than the majority of U.S. distributors are capable of providing (1). Consequently, we are of the opinion that the availability of technical information centers and laboratories is vital for the U.S. fastener industry.

With access to a good fastening laboratory, marketing can be done from a position of strength. Confidence is built up between the fastener users and the fastener manufacturer; also, and of equal importance, confidence between the raw material supplier and the fastener manufacture will be engendered.

8.2 FUNCTIONS OF A FASTENER LABORATORY

Nine variables are associated with screw-fastening evaluation and testing, regardless of whether the material is plastic or metal:

1. Type of fastener
2. Head style
3. Fastener finish
4. Diameter of fastener
5. Length of fastener
6. Proper pilot-hole size, including taper
7. Boss OD
8. Evaluation of fastening site in the selected material
9. Type and speed of assembly equipment

Some of these will already be known to the designer, engineer, or inquirer because they are dictated by product design.

8.2.1 Fastener Research : A Cooperative Venture

There really are no fixed rules or standards that apply to the subject of fastening in plastics, other than actual laboratory testing. Therefore it is absolutely vital that determination of fastener usage in plastics be a cooperative venture between the design department of the parts manufacturer and the fastener laboratory. In examining potential applications jointly, consultations will establish the optimum relationship of screw and plastic material at the fastening site.

8.2.2 Subjects to Investigate

The following areas should be covered:

1. The general capability of a particular plastic to be used successfully with the fastener.
2. Fastening-site design: length of engagement, boss wall thickness,
hole, and screw diameter.
3. Drive and seating torque values that should be used on the
assembly line and the best surface finish for the screw.
4. Selection of the proper geometric screw design for the particular
application.

As for the laboratory, knowing the material, and best of all getting
a sample of the actual material to work with is of utmost importance, as
results will generally differ from supplier to supplier. We have seen
examples of different results in the preceding chapter. The best possible action is to supply the fastener laboratory with an actual example
of the part with established fastening sites, or a mockup if an example
is not available. Also the laboratory will need strips or blocks of the
resin producer's material.

With the actual material to work with, the fastener laboratory should
be able to supply recorded data on:

1. Optimum hole size for each type and size of fastener at its
maximum drive-strip ratio
2. Recommended thread engagement
3. Recommended tightening torque
4. Pull-out force

The report will be in terms of drive torque, prevailing off torque, fail
or strip torque, pull-out tension, and fail-drive differential, and perhaps, depending on the application, clamp load, load relaxation, or
resistance to creep.

8.3 DEFINITIONS OF TERMS USED

Drive torque. In the case of a thread-forming or thread-rolling
screw, drive torque is the torsional effort required to overcome the
resistance encountered in entering the pilot hole, the required length
of engagement in the assembly nut member. It is made up of two principal components: friction (generally the major component) and thread-
forming action (material displacement).

The high friction content of drive torque is attributed to the heat
generated in driving, the zero-clearance interference fit between screw
and internal thread and the resiliency (or elastic memory) of the plastic,
which tends to close back against the screw.

Prevailing off torque. This is the torque required to back the fastener out of the hole without any influence from seating it. It is an indication of resistance to vibrational loosening. Complete laboratory reports sometimes show the results of first and fifth prevailing off torques. The first prevailing off torque is often higher than the initial drive torque. Also in plastics, if frictionally created heat is allowed to dissipate at the end of each cycle, there should be a lowering of the prevailing torque through the five readings.

Fail torque. As a result of continued application of drive torque, failure occurs at the fastening site either because the screw breaks or the material is not strong enough to withstand the applied torque force and the previously created threads strip. Failure can occur from either screw breakage or thread-stripping, and the term "fail torque" is often applied to either. Failure can also come about from stripping of the drive recess or slot.

Fail tension. This is the highest clamp load achieved when the assembly is "torqued" to the fail point.

Fail-drive or fail-strip ratio. Fail or strip torque divided by drive torque, and written 3:1, 3.8:1, etc. Fail-strip ratios well in excess of 3:1 are desirable for high-speed assembly operations.
Remember, we are discussing plastics. In metals, designers strive for a fail-drive ratio of at least 3 to 1 in judging suitability for use in high-production automotive-assembly conditions. But in plastics, because of the low magnitude of torque usually achieved, an even higher ratio is desirable.

Fail-drive differential. In plastics, a high ratio is good, but it does not tell the whole story. With plastics, it is better to talk about the *differential* between drive and fail torques. Again, a high value is good. It is desirable because it assures the engineers on the assembly line of a wider opportunity to choose a driving torque that will drive the screws efficiently without stripping the plastic or breaking the screws. When a wide differential is available, it is easier to set up and adjust automatic screwdrivers to cover the variances which occur in working with products made from plastics.
Because of the low torque values usually achieved in plastics, differential is more indicative of the torque that can be obtained.

8.4 DETERMINING OPTIMUM HOLE SIZE

A great many times, perhaps most of the time, it will not be necessary to determine optimum hole size from scratch. Parameters are usually already established and the laboratory technician or engineer will be trying to rectify a problem some manufacturer got into by not contacting the laboratory soon enough. However, when the opportunity does present itself on a brand new application, the laboratory engineer will probably follow a course of action similar to the following:

1. Determine the general properties of the plastic as to its degree of stiffness of resiliency. This will be done either through experience, collected file information about the material, or direct contact with the resin producer.
2. Determine the screw style to use, if the company manufactures more than one type. In these days of computers, laboratories should be able to keep a cross-referenced list of laboratory studies by type of screw, size, material, and so forth to make this a quick task.
3. Refer to technical data sheets which indicate the range of pilot holes to be used in plastics. Select an initial, test-hole size consistent with the degree of stiffness or elasticity of the material. Softer, more ductile plastics should be able to take the smaller holes with more dense materials requiring larger holes.
4. Experiment with hole sizes slightly larger and smaller than the initial selection until the highest pullout is achieved for the torque desired by the parts manufacturer. (In most cases, the parts manufacturer will be working with power screwdrivers with a fixed torque level.)

Information on required number and nominal size of fasteners to be used ought to be supplied to the laboratory by the parts manufacturer. The part designer will have to calculate the forces the assembly must withstand under use, over the projected lifespan of the part. The designer will use standard engineering formulas available in strength of materials and machine design handbooks to obtain the data.

However, it should be noted that many joint-fastening formulas which are considered tried and true with metals, do not apply to the viscoelastic plastics.

It will be helpful to the fastener laboratory engineer to be able to work with charts for the fastener which show percentages of thread-engagement depth into the side of the hole, for various hole sizes. Such a chart can easily be developed knowing pitch diameter, included thread angles, and mean major and minor diameters. We are talking about drilled holes, of course, unless the pilot-hole size has already been determined and actual taper-cored holes are available for testing. If the boss size has already been determined, our choice will be affected by the desirability of keeping the ratio of boss OD to pilot hole between 2 to 3:1. Since we are determining which screw to use in a taper-cored hole, the drilled-hole diameter will be considered to be at the mean location.

Length of engagement. This is usually a fixed length, for example, when working with through holes. With blind holes, there will be some leeway, but for optimum engagement length we should talk in terms of 2½ to 3 screw diameters.

Optimum conditions for assembly will have been achieved when screws and application site fail at approximately equal degree; that is, when the screw breaks about 50% of the time and strips about 50% of the time at the same drive torque. Recessed drive screws often camout and strip threads at close to the same values. If camout is a problem, the use of well-designed hex heads and/or Torx®, Supadriv®, or Pozi-Driv®* recesses (all of which reduce camout probability) should support higher torque levels.

RPM of drivers. The laboratory with which we are most familiar does most of its testing at speeds of either 250 or 450 rpm; of course, in any one test the same rpm is used throughout the test.

It would be useful to the fastener industry if a laboratory would investigate and publish the effects of driving at various speeds in different materials.

8.5 LABORATORY EQUIPMENT IN ANALYSIS OF PLASTICS APPLICATIONS

The tools of the trade used by one excellently supplied laboratory with which we are familiar include among other laboratory equipment:

1. Electronic torque transducer with 200 lb-in. capacity (Fig. 8.1)
2. GSE torque-tension unit with integrated load cell for obtaining tension such as seating, fail tension, and pullout (Fig. 8.2)
3. Dual-channel bridge amplifier/recorder for graphically plotting torque-tension curves (Fig. 8.3)

If desired, the units could also be set up to obtain both the torque required to overcome friction and the torque required to produce the thread in the material. Besides adequate testing equipment, other advisable equipment might be temperature and humidity controls.

---

*Pozi-Driv® is a registered trademark of Phillips Screw Co. Supadriv® is a registered trademark of GKN Fasteners Limited. Torx® is a registered trademark of Camcar-Textron.

FIG. 8.1   Laboratory setup for reading torque on application
discussed in Chapter 7, Sec. 7.6, with spindle-mounted GSE electronic torque transducer.

FIG 8.2   Loading material test block into GSE torque-tension unit.

FIG. 8.3 Laboratory setup for obtaining drive torque, prevailing
on torque, prevailing off torque, clamp-load torque, plus seating
torque and tension. Hydraulic equipment for axially pulling screws
in tension is not shown. Dual-channel, bridge amplifier/recorder by
Gould, Inc., is at right.

8.6 LABORATORY STUDY EXAMPLES

Over several years, a series of interrelated laboratory studies were
done by a well-known fastener company for a customer in the electronics communication field. Initially the customer was developing new
fastener parameters for a metal electronics housing framework. During
the course of the studies, the customer eliminated use of the metal
framework, tried glass fiber, and eventually settled on use of foam
polycarbonate.

These studies not only illustrate what we have been discussing, but
also indicate that the values achieved in metals can now be approached
with plastics.

Previous to the initial study, which we will label study A, the customer had been assembling the structure using no. 4-40 stainless steel
screws with crowned copper washers in tapped holes. The customer's
engineers approached the fastener manufacturer to determine whether
they could achieve similar fastening values and also eliminate the tapping operation by use of the fastener manufacturer's special-design
thread-rolling screws for metal.

The customer's design engineers reported that they had been getting
a minimum of 150 lb clamp load at 4 lb-in. seating torque. The 150 lb
clamp load was required to cause the washer to flex and to be functional.

Perhaps the flex was required because of an electrical connector. The study did not elaborate. But in any case, as long as the washer remained functional a higher clamp would have been permissable. In study A, preliminary tests were done with both the machine screws and with the special-design thread-rolling screws. It was determined that at 4 lb-in. seating torque a consistent 150 lb clamp load could not be achieved. It was also found that it was not possible to seat the special-design thread rollers at 4 lb-in. using a pneumatic power screwdriver because the required thread-forming torque was slightly higher.

However, at 6 lb-in. seating torque, consistent seating was achieved without causing washer damage. This produced well over 150 lb clamp load, and the setting was suitable for power-driving.

Torque-tension tests were performed exactly as previously described and the charts, *similar* to the one shown in Fig. 8.4 were analyzed. The figures were then applied to a computer program which statistically analyzed data enabling the plotting of a straight-line graph presentation of the torque-tension relationship (Fig. 8.5).

In study B, made some months later, tests were run for the customer with the crowned copper washer under the head of the screws and bearing against a glass fiber circuit board material. The nut member consisted of .093 in. thick cold rolled steel. A series of six tests were conducted with both painted steel and nickel-plated steel nut members in three different hole diameters (.098, .100, and .102 in.). Tests results follow in nickel-plated nuts with .100 in. holes:

Screw: No. 4-40 × 5/8 in. slotted pan head, special-design steel thread-rolling screws for metal, zinc dichromate finish

Torque values (lb-in.)		Tension values (lb)			
Drive torque	Seating torque	Seating tension	Fail torque	Fail tension	Type of failure
4.3	6	158	25.2	570	Screw broke
4.1	6	180	22.4	555	Screw broke
4.4	6	185	24.8	465	Screw broke
3.9	6	225	17.0	420	Screw broke
3.8	6	188	16.0	600	Hole stripped
4.1	6	158	17.6	390	Hole stripped
4.1	6	180	20.0	390	Screw broke
Avg. 4.1		182	20.4	484	

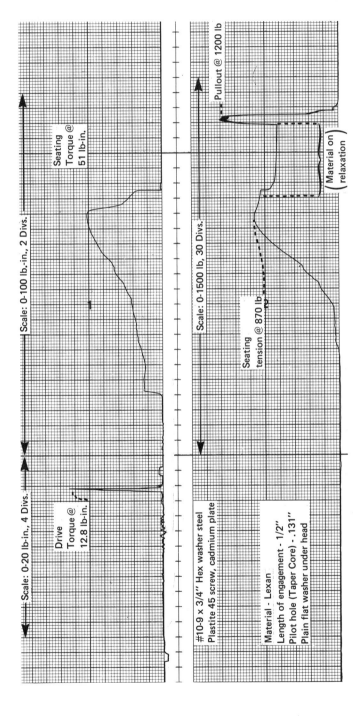

FIG. 8.4   Typical torque-tension graph from dual-channel recorder. In the chart the material's initial load relaxation shows up clearly. (Torque values, lb-in.; seating tension, lb.)

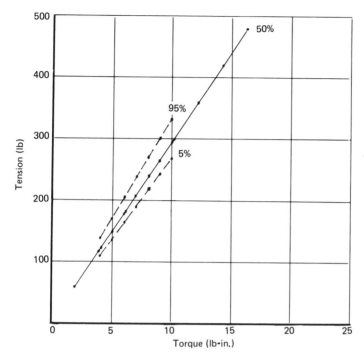

FIG 8.5   Torque-tension plot, based on computer analysis of data in study A. No. 4-40 × 1/2 in. slotted pan head steel trilobular screw with zinc and dichromate finish in .093 in. thick CRS nut (RB 85-95) with .098 in. pilot hole.

Study C was conducted about 9 months later, when the customer again contacted the laboratory, having developed a product of like design in structural foam polycarbonate. The test arrangement was very similar, but the washers were 1-in. square Griptite design beryllium copper and the plastic test samples were .250 in. thick, not .093 in. thick as with the cold rolled steel nut members.

The customer's engineers felt that 200-250 lb clamp load was desirable to secure the assembly. They did not want to fracture the washer and they did want from 1/2 to 5/8 in. engagement length, as pullout and ultimate assembly torque was also of concern.

No. 4-20 × 7/8 in. slotted hex washer head 60°, single-lead special-design thread-rolling screws for plastic with cadmium finish were used in the structural foam, polycarbonate test blocks at a 1/2 to 9/16 in. engagement. All tests were run using the equipment previously covered and in a similar manner. Test results follow:

Torque Values: lb-in. Tension and Pull-out Values: lb.
Test No. 1

Drive torque	Torque at tension 200 lb	250 lb	Fail torque[a]	Fail tension	Tensile pullout	Type of failure
4.5	9.5	11.5	18.0	540	—	Camout[a]
4.1	8.0	10.1	18.0	525	—	Camout[a]
3.9	9.6	11.4	15.3	390	810	Camout[a]
4.8	6.6	11.1	16.2	390	750	Camout[a]
4.2	9.0	10.5	21.3	660	—	Hole stripped
4.0	9.9	10.5	22.8	615	945	Camout[a]
3.0	7.8	9.1	18.9	600	885	Camout[a]
3.9	8.1	11.1	21.9	675	960	Camout[a]
Avg. 4.1	8.6	10.7	19.1[a]	549	870	
Sigma .5	1.1	.69	2.7	11	89	

[a]In most cases the assembly did not actually fail, as failure occurred
due to camout. For this reason, pull-out tests were also performed
and assemblies actually failed by hole strippage. Fail torque average
is based on values where camout occurred. However, it demonstrates
the high levels at which failure occurs. Normal seating torque ranged
from 8 to 11 lb-in.

In test no. 1, the screws were driven into the plastic at 300 rpm using
a hand-held pneumatic driver. Drive torque, failure torque, and
tension, as well as torque-tension curves were graphically recorded.
The torque required to achieve 200 and 250 lb was also recorded.

In test no. 2 (below), the torque tension results of test no. 1 were
studied and it was decided that 200 lb clamp load would be the goal.
It was felt that 9-11 lb-in. torque would be required to achieve the
desired clamp load.

Screws were driven at 300 rpm using the hand-held, pneumatic
driver and recording maximum drive torque. A hand-held driver was
then used to accurately seat screws at 200 lb clamp load. On all assemblies except one, no significant loss of load was noted. Low loss of load
was attributed to the low creep inherent with polycarbonate and to
screw design. After each screw was seated, the assembly was pulled
apart and pull-out load recorded.

In this test, polycarbonate materials from two different resin producers were analyzed:

Test No. 2

Nut member supplier	Drive torque	Torque at 200 lb	Tensile pullout	Type of failure
A	5.6	10.5	775	Hole stripped
B	3.8	9.5	750	Hole stripped
A	5.6	10.4	800	Hole stripped
B	4.3	9.9	738	Hole stripped
A	3.6	9.5	700	Hole stripped
B	4.3	11.1	813	Hole stripped
—	5.4	—	750	Hole stripped
—	6.1	—	850	Hole stripped
—	4.5	—	850	Hole stripped
Avg.	4.8	10.2	781	
Sigma	.89	.63	52	
Avg. A	4.9		758	
B	4.1		767	

Test no. 3 was performed to eliminate possible concern over loss of assembly strength after multiple in and out tightening and loosening cycles. A screw was seated at 200 lb, recording required torque, then broken away and reseated. After 6 cycles the assembly was pulled apart, recording pull-out strength.

Results were: seating torque = 10.4, 10.4, 10.4, 10.4, 10.8, 11.2 lb-in.; pullout - 825 lb.

In this series of tests it is interesting to note that the drive torques and tensile pull-out test results for two different resin suppliers' products were quite similar. The material tested was in the form of test blocks that would fit into the laboratory fixtures. However, the blocks came from resin producers. We are not talking about finished molded pieces, which are not so easy to test in such depth without designing special fixtures. We should also point out that relatively similar fastening results can be achieved for both metals and structural foam polycarbonates when the fastening site can be designed for it.

With the types of sophisticated electronic equipment now available, laboratory techniques have left behind the static fastener testing of 20 years ago. A simple torque wrench cannot provide the dynamic type of testing now possible, testing which can essentially duplicate field conditions. Fastener organizations which can provide the sort of information discussed in this chapter will be the ones that have the most opportunity to command a large market share, no matter in what materials the fasteners are to be used.

The availability of such information frees the screw or insert manufacturer from hand-to-mouth conversion of metal into mere commodities. With the laboratory, it is now possible to embark upon a true marketing campaign to provide successful "fastenings." Successful fastening are the key to long-term profitability for the fastener manufacturer, lower in-place cost for the finished product manufacturer, and satisfaction for the user of the assembled product.

REFERENCE

1. F. Akstens, "Swiss Metric Fastener Distributor Opens U.S. Facility," *Fastener Technology*, 5:30-31 (October 1982).

9
POWER SCREWDRIVING EQUIPMENT

This chapter will cover air-powered and electrically powered screw-driving equipment. The focus is primarily on portable equipment, but we will also briefly touch on bench models and the fastener quality levels that they require. Operating speeds will be considered, as well as accuracy levels, available torque ranges, and the categorizing of applications by whether the joints are hard or soft pull-up (draw).

Our opinion regarding tool cost is the same here as for fasteners; it is not the immediate cost of the equipment which is important. It is the overall cost of the fastening operation itself that counts.

Some equipment manufacturers are listed. We intend to cover those that we feel provide their technology as part of their product line, the ones we believe maintain "product managers" in the field, able to give advice on fastening applications. There may be others that have been overlooked.

In connection with equipment and the need of cooperation from equipment manufacturers, it should be immediately stated that there can be no substitute for trying out the tool recommendation when you are getting set up for an application. Try it before you buy it! It is merely common sense to have this attitude toward the equipment as well as testing out the screw recommendation from the laboratory.

Although descriptions of portable power screwdrivers are provided in some detail so that a good understanding will exist, we are actually more concerned about how their use affects the assembly of helically driven screws.

The power screwdrivers mentioned in this chapter are designed to shut off or declutch at a predetermined torque achievement level. They are used on assembly-line applications where precise torque level control is mandatory or desirable, and that situation covers most assembly operations in most industries.

SELECTION OF THE DRIVER TOOL

Before proper determination of which torque level control driver will be most satisfactory, the type of joint should be identified.

9.1 EXPLANATION OF "SOFT" AND "HARD" DRAW, SLAM, OR PULL-UP

In the industry it is customary to categorize all applications as "hard" or "soft." The words refer to two kinds of joints and the generated clamp load being achieved during the joining. This is to say, they designate whether the clamp load is either instantaneously achieved or perhaps gradually achieved because of some absorption in the middle of the joint.

A "soft" joint is one that never stops being soft, similar to driving a wood screw. It can be like the drawing together of a sandwich of materials such as when assembling the leaf springs of a truck with nuts and bolts. Indeed, there is generally either a gasket, or something which acts like a gasket, between the two materials that the fastener is joining.

Such an application is illustrated by the electric plug (Fig. 9.1) intended to eliminate any exterior sparks. The design of the plug is such and the fit of the segments is so close that the assembly complies with the soft joint described above.

Achievement of clamp load and desired torque shut off with soft pull-up can be diagrammed as follows:

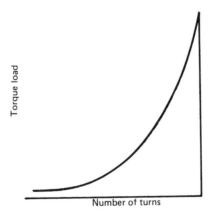

"Hard" draw or slam described the situation where there is either a free-running condition of screw entry into the hole (or a constant level of torque) until just before the screwhead seats, when there is a sudden buildup of resistance. This occurs usually with the harder, stiffer plastics. When the head seats, it does so instantly. An application of this type is shown in Fig. 9.2.

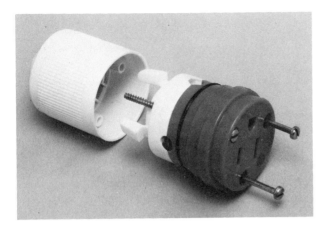

FIG. 9.1 Soft pull-up application in which there is a gradual buildup
of resistance until the screw is fully seated. Material is Nylon.

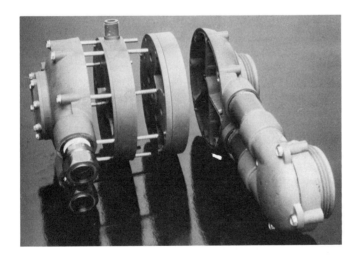

FIG. 9.2 In the water softener application described in Sec. 7.6 and
manufactured from Noryl with 30% glass-fiber reinforcement, there is a
sudden buildup of torque just before the head seats.

The achievement of clamp load under hard slam conditions is indicated
by the following diagram:

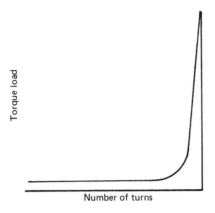

9.2 SCREWDRIVER CONFIGURATION

There are three main types: (1) straight body, mostly used in vertical
applications, (2) pistol grip, generally for horizontal insertions, and
(3) angle drivers (Figs. 9.3a-d).

9.2.1 Straight Body: Lever or Push Control

This design is sometimes referred to as a pencil driver. It is a narrow,
straight-body unit ideal for small-size fasteners. The configuration
can be used in fastening applications up to 100-110 lb-in. torque, considerably more torque than is generally required for use in plastics.

In the straight-body type, motor and drive are both on the same vertical axis; the driver is held by the operator like a pencil. Rotation is
activated by a lever which extends parallel to the body or by a "push-
to-start" feature on the bit. This type of tool is usually attached above
the assembly area and pulled down into position by the operator. The
long, slender body is especially adaptable to small assemblies where
space is at a premium. Some suppliers have adaptors which can be used
to convert a straight-body driver to the pistol-grip style. Either air-
powered or electrical in-line drivers can be adapted for insertion of
coiled wire inserts by means of a special prewinding tool and mandrel
as described in Chap. 10, Sec. 10.5.1.

9.2.2 Pistol Grip

This design looks just as the name implies and because of the shape can
transmit greater torque than the straight-body type. However, pistol-
grip units are often used for small size screws in the No. 2-12 diameter
range. This is the general area we are concerned about in fastening

(a)

(b)

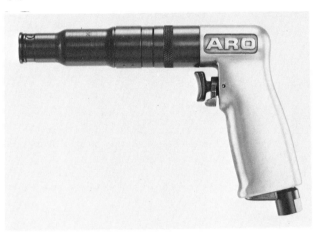

(c)

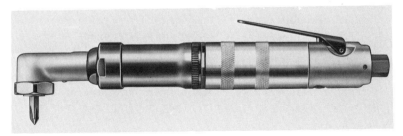

(d)

FIG. 9.3 (a) Straight-line driver, level actuated; (b) straight-line driver with push-to-start feature; (c) pistol grip; (d) right-angle driver. (Courtesy of The ARO Corporation, Bryan, OH.)

plastics, although sometimes there can be applications using fasteners
up to 5/16 in. diameters.

Air tool suppliers often provide pistol-grip tools with torque reaction
bars that allow these tools to be used with larger diameters in metals.
However, in the plastics the use of torque-reaction bars are not generally necessary.

9.2.3 Right–Angle Drivers

Right-angle drivers have drive-spindles at a 90 degree angle to the
combination handle and body motor housing. Some 45 degree angle
units are also sold. The primary use of the angle drivers is in large-
capacity applications where the long-length handle-body motor housing
can be used to withstand torque reaction. However, in the smaller applications the configuration could be handy for cramped assembly conditions. Right-angle tools can be obtained in soft-draw design, which is
good for many insertions in plastic assemblies. Right-angle tools with
an air shutoff feature are also available for torque ranges from about
10 to 150 lb-in.

9.3 ELECTRIC OR PNEUMATIC DRIVERS?

We mentioned electrical screwdrivers at the very beginning of the chapter. Currently, there do not seem to be many manufacturers in the
United States, and the majority of interest appears to be in the pencil-
type units for very small or relatively small screws. As we recall, our
Japanese friends in the fastener field began talking about these small
units about 10 years ago. We understand that they have become quite
popular in Japan.

Although concentration of interest may be in the small units, electric
screwdrivers are available in all three driver configurations. We were
discussing recently why you do not see more of the electric pistol type.
Our conclusions are that in the past the electric equipment for the higher torque values probably compared unfavorably with pneumatic units
with regard to weight and bulk. This may have prejudiced users.

It is a different story, however, with the small, pencil-type equipment,
which appears to be ideal for very small screws up to about no. 4 diameter where torque values run up to about 5 lb-in. They have also been
available for fasteners up to no. 10-24 for use at torque levels up to
about 20 lb-in. and therefore are suitable for many applications in
plastics.

The electric equipment is available in the United States through at
least several sources. Units manufactured by Hios in Japan are sold by
their distributor Jergens, Inc. in Cleveland. The Foredom Electric
Company in Bethel, Connecticut manufactures a reversible small, in-
line driver. It runs on AC current at 1650 rpm, for torques ranging
up to 5 lb-in. Although their catalog does not say so, they can rewire
the unit to provide 7-8 lb-in. torque. Weber also has an electric in-
line driver, but it is not listed in their current catalog. Recently, the

FIG. 9.4 In-line, DC driver with adjustable 20/30 V transformer.
(Courtesy of Jergens, Inc., Cleveland, OH.)

available torque range has been extended to 30 lb-in. with the introduction of some new models by Desoutter.

The in-line drivers sold by Jergens (Fig. 9.4) operate on DC current
and have transformers which operate at either 20 VDC or 30 VDC, in
either case at moderate rpm. The medium-range, in-line units operate
at reasonably wide torque ranges adjustable through external nuts.

Electric screwdrivers have many good features, not the least of which
is low noise. Furthermore, because there is no discharge of air as with
pneumatic drivers they are valuable for use in "clean rooms." There
is a market for electric drivers, wherever the electronics industry is
located, and more manufacturers will enter this field, soon.

Nevertheless, the use of air-powered screwdrivers is the most prev-
alent, and we will cover these next.

9.4 DRIVING METHODS FOR AIR-POWERED PORTABLE
SCREWDRIVERS

With pneumatic equipment, if you are going to be involved in a multi-
driver installation, regardless of the equipment configuration, you
should aim at a control situation that is analogous to that of modern
television sets, where picture size is stabilized, even though the voltage

may fluctuate. It is desirable to be able to maintain constant torque-level control in spite of fluctuations in air supply.

For the sake of simplicity, the driving methods of air-powered portable screwdrivers have been categorized as either having clutch control, or not having clutch control. There are two main types of equipment: (1) direct drive-air stall and (2) automatic or mechanically adjustable clutch drive control.

9.4.1   Direct Drive—Air Stall

The control method used with this equipment is called "stalled torque control." In this system, the motor spindle is connected directly to the driving rotor without any clutch. It is adjustable only by control of the air supply. When desired torque has been achieved, the motor stalls. Air-line pressure is controlled through a pressure regulator. In the stalled condition, the motor actually stops, even though the air continues to flow. This condition would of course burn out the usual electrically operated gun. This type of unit has no air-shutoff feature.

Cost. With direct drive, motor power is transmitted to the driving action at the highest efficiency; whereas with clutches only 40-60% can be transmitted. Therefore, a smaller motor can be used with direct-drive systems, and they should cost less.

Sizes of fasteners that can be used with direct drive—air stall. Remember that we are talking plastics. This generally means that 1/4 in. is the top diameter. The torsional reaction of direct-drive equipment is not a real factor in plastics applications; whereas it could really be intolerable to the operator in some metal applications.

Direct-drive—air stall units can be used for either hard slam or soft draw situations. In the latter condition they may have to be set to run rather slowly (with small screws) to prevent "camout," where the driver blade slips out of the slot or recess. However, we feel that in plastics, "camout" should not be a serious problem. Provided driver bits are in good shape, proper screw selection and the use of correct hole size should eliminate the problem. Here's where the fastener laboratory will help.

Achievable accuracy with direct drive—air stall screwdrivers. The ability to have accurate settings is more important with plastics than with metals. In plastics we are generally dealing with a lower magnitude of numbers than in metals. You do not have the higher strip-to-drive ratios or the higher differentials between strip (fail) and drive torques that you find with metals. Furthermore, with plastic materials there can be more variation in torque values because of the wide range of differences in material properties.

However, as long as you have a constant air supply, direct-drive—air stall equipment is accurate enough for general applications and because of the lower initial cost should certainly be considered. They can be all that is required for laboratories or small shops. In larger setups you can use them also, provided air regulators are used and each gun is set to stall below the maximum airline pressure drop.

There are actually no set parameters for minimum air pressure; in a given situation, depending on the plastics application, you might get satisfactory settings at a constant psi of 40 or below. It entirely depends on the kind of application being dealt with. Again we should emphasize that there is no substitute for trying out the system. That is why it is important to deal with equipment suppliers who have the technical expertise, the personnel, and the desire to work with you. The driver manufacturers will probably steer you in the direction of clutch control units, especially for use with small fasteners.

9.4.2 Automatic or Mechanically Adjustable Clutch-Controlled Screwdrivers

Clutch-controlled power drivers are found in the field much more frequently than the direct-drive—air stall type. The reason is one of convenience; the fact that they are adjustable within ranges of torque is an important consideration since they may be used more easily on a wider variety of applications. Additionally, with clutch control, operator technique is not so important, and less training is required.

Clutch-control power drivers are currently available in four major types only, although one can be easily confused by the habit of some suppliers to consider variants from these to be a separate type. The four types are: (1) direct clutch, (2) positive clutch, (3) adjustable clutch, and (4) adjustable clutch with air shut-off.

Direct clutch drivers. These are direct-drive-air stall types with an added clutch which is disengaged until the operator exerts end pressure (Fig. 9.5a). They have all the advantages and disadvantages of the standard direct-drive system; the motor operates, but the gun will not drive without a push being given. This system is often recommended because of the ease of engaging driver bit with the driving system on the screw. The "push-to-start" feature is usually combined with adjustable clutches.

Positive clutch drivers. These are similar to direct clutch types, except that the clutch operates as a direct clutch until the torque is high enough to cause the clutch jaws to ratchet. This allows the creation of added torque build-up until the operator disengages the tool. The system is valuable for situations where they may be variation in material densities between two plastics or if density variation could occur within within an assembly in the manner that it can occur within wood, which may have knots. With positive drive, operator technique is required. Figure 9.5b illustrates a positive clutch drive unit.

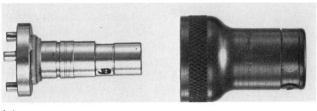

(a)

(b)

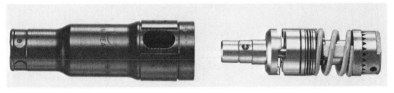

(c)

(d)

FIG. 9.5   (a) Direct clutch driving mechanism; (b) positive clutch driving mechanism; (c) adjustable clutch driving mechanism without air shut-off; (d) adjustable clutch driving mechanism with air shut-off. (Courtesy of The ARO Corporation, Bryan, OH.)

Adjustable clutch drivers without air shut-off. For soft-draw or hard slam situations, adjustable clutch drivers may be supplied with ratchet or friction type clutches with adjustable spring tension; the best types have some variation of ball-and-spring configuration. The air continues to flow with these models; only the clutch disengages from rotation (Fig. 9.5c).

Torque, shut-off-level adjustment is via an internal spring which holds a fixed clutch and a sliding clutch member together. Adjustment is by trial and error often on a nut located inside the housing. In this case, the removal of the cover is required. Other designs have an external, sliding ring or some other external device that can be adjusted with a tool or sometimes merely with the fingers.

Adjustable clutch drivers with air shut-off. This kind of driver is generally similar in control configuration to the adjustable ball-and-spring type mentioned above. However, there is usually a central spindle or push rod running through the clutch assembly which acts to immediately turn off the air supply on achievement of the desired torque level. There are designs which do not declutch, where the push rod shuts off a starting valve. In this case, residual inertia is absorbed through the main spring (Fig. 9.5d).

Since the use of air is in itself relatively high cost, the use of air shut-off drivers appear to be a good investment, in spite of the initial higher tool purchase price.

Air shut-off systems reduce torque reaction to a very low level, thereby cutting down on operator fatigue. The system is said to be quieter because of the shut-off feature.

9.5 MAKING THE CHOICE OF TYPE OF SCREWDRIVER

Selection of the tool configuration (straight-body, pistol-grip, or right-angle driver) as described under Sec. 9.2.1-9.2.3 will be dependent on a review of the assembly setup and the conditions of the assembly.

Selection of the best torque control method can be achieved by using the following chart. An analysis of the best method should be made after classification of the joint as being either "hard" or "soft." Final selection will also be dependent on desired speed and fastener as well as the material.

We are fully aware that some users will say "I'm going to use what's available in-house." That can be done, too, provided in-house equipment will give the lowest in-place fastening cost—a consideration which can encompass many factors, including anticipated run length. Fastener design recommendation should consider the equipment to be used and its torque control range.

In most cases, it will pay to examine the ideal selection, which can be "roughed out," as follows:

Usage Recommendation—Various Assembly Conditions in Plastic

Application type	Direct drive—air stall			Adj. clutch (ball and spring)	
	No clutch	Push-to-start feature	Positive clutch	Without air shut-off	With air shut-off
Free or easy running, *followed by*:					
Hard slam	Marginal	Marginal	Not recommended	Very good	Excellent
Soft draw (gradual resistance increase)	Marginal	Good	Marginal	Very good	Recommended on low to intermediate torques only
Constant, moderate resistance, *followed by*:					
Hard slam	Marginal	Good	Marginal	Good	Usually not recommended but try it!
Soft draw (gradual resistance increase)	Marginal	Good	Marginal	Good	Usually not recommended but try it!

9.6 OPERATING SPEEDS

The operating speed of air-powered screwdrivers is an important factor
to be reviewed when selecting a unit, after decision has been made about
desired torque settings.

What manufacturers of air-powered screwdrivers list in their catalogs
is *free-running speed*, not the speed under load, which will be lower.
As a general rule of thumb, we are told that the maximum efficiency of
air-powered screwdrivers is attained at about 65% of their listed free-
running speed. That would seem to be approximately the speed we
should try to obtain.

When it comes to driving-speed recommendations, we tend to be on
the conservative side. In the absence of a specific study or experience
with a given plastic, it is our belief that high rpm should be avoided
because of the frictional heat produced in the plastic during the driving
operation. We are of the opinion that high heat may cause an expansion
and or melting of the plastic and consequent lower clamp load, especially with those of lower melt levels. Indeed, most of the laboratory
studies with which we are familiar are run *under torque load* (not free-
running) at speeds of 400-500 rpm, as originally measured by a strobe-
light timing gun.

At the more moderate speeds, repetitive torque control is more easily
attained. This is due to decrease in inertia and consequent reduction
in the slam effect. Indeed, an experienced individual has said, "Use
the very slowest possible speed any time that operator technique is
involved." And *with very small screws* (say, sizes #2 to #4) this recommendation may have special weight because the available differential
ranges between drive and fail torques are numerically smaller with
these sizes.

We are told that with small screws (and an increasing amount of smaller
screws is being sold), the most popular free-running speed request is
for 900 rpm and the next most popular call is for 1500 rpm. The latter
seems on the edge of being too rapid for use in plastics, where *under
torque load* we have been given recommendations not to exceed 800 to
1000 rpm.

Final recommendations should be obtained by analysis that considers
hole size and depth, material type and thickness, and other factors used
in determining torque values. Once the joint has been properly evalu-
ated and torque level decided on, consultation with the screwdriver
manufacturer's product manager should result in prompt advice on the
speed and driver to use for the particular assembly.

9.7 ACCURACY

The consistency of the applied torque will be affected by the speed of
the driver. At higher speeds the power tool does not stop as quickly

because of built-up inertia. Since the highest torque is arrived at, just before the torque reaction effect happens, air-stall and shut-off reaction at lower speeds should be more consistent (1). In selecting power screwdrivers, considerable attention should be paid to the method used in stopping the built-up inertia.

The degree of repeatible torque control accuracy with *electric* drivers is probably higher than with pneumatic units. We not the Jergens' catalog refers to an achievable accuracy of ±1-1½%. We also note that only a few manufacturers of pneumatic equipment seem to cover this subject with specific percentages of accuracy. A worker with over 25 years experience in the field seemed to feel comfortable with ±5 to 8%. This seems like a convincing reason to shoot for wide differential between drive and strip torques on fasteners used in plastics. We will get back to the subject of repeatable accuracy in Sec. 9.9.

9.8 TORQUE LEVEL RANGES

When equipment manufacturers list torque level information, they catagorize the ranges under the soft draw and hard slam screw entry conditions that we previously discussed. Soft draw ranges *may* show maximum and minimums. They will apply only to clutch-controlled equipment. Hard slam applications are generally free-running until immediately prior to screw head seating, so the listings would show only maximums. Direct-drive equipment also would show only maximums, since operator technique is involved.

A few typical ranges are as follows:

| | | | | | Torque (lb-in.) | | |
| | | | | | Soft | | Hard |
Model	rpm	Throttle	Clutch	Screw size	Min.	Max.	Max.
A	1100	Push	Adjustable	#2-5	2	5	8
B	900	Lever	Direct drive	#6-1/4	*	80	—
C	600	Lever	Positive	#4-10	*	50	—
D	750	Pistol	Adjustable	#2-8	3	16	18
E	750	Pistol	Positive	#4-10	*	50	—
F	750	Pistol	Direct drive	#4-10	*	50	—

*Controlled by operator technique.

The speeds shown here are quite a bit lower than are usually found listed in the driver manufacturers' catalog. However, we are dealing with plastics, usually fastened with screws having half as many threads per inch as machine screws, or even less than half in the case of the special-design thread-rolling screws for plastics. Naturally, the higher helix angle on the threads of these screws means that little time will be sacrificed.

9.9 TORQUE CONTROL AND SIZE SELECTION IN PLASTIC APPLICATIONS

Although the subject of torque is covered in other chapters in more detail, the reader should be reminded that with plastics a major problem, perhaps we should say *the* major problem, especially with small screws, is to get the drive and fail torques up high enough for effective assembly-line threading to take place. We are particularly desirous of obtaining a wide differential between drive and fail torques, so that driver adjustment is sure and easy.

In setting up the equipment, one should strive to have the torque control setting as close as possible to the middle to three-quarters of the available range. At least one manufacturer recommends that the maximum torque value of the tool should be a minimum of 20% greater than the clutch setting. This will allow for a drop in air pressure and possible loss in tool efficiency from misuse.

Screw selection will enter into the picture; we refer the reader to Chapter 6, Sec. 6.4.1, for a synopsis of steps to consider for the correction of difficulties.

With metals it is possible to tighten bolts into their elastic ranges. By doing so, we produce a condition of maximum preload (applied tension) which offers increased fatigue resistance and greater reliability(2). But with plastics this is not usually possible because of the lower strength of the materials, greater elasticity of the plastics, and so forth.

While the setting accuracy of torque-control air-powered drivers has been improved and is reasonably accurate, they could not be used to tighten fasteners into their elastic ranges, since after the plastic starts to yield, the torque remains nearly constant, so that the driver would continue to rotate until rupture occurs (2).

Consequently, torque limits are going to have to be determined by some other means. In practice, the torque setting should be based on and set lower than the fastener's "ultimate tensile strength," as recommended by the fastener manufacturer. His recommendations will involve a good many other factors, including the plastic material's coefficient of friction, whether he is suggesting the use of "torque robbers" (serrations under the head of the screw), hole size, type of fastener, and so forth.

Because of the large friction component of torque (say, 90%), only 10% of the torque may be available for applying tension. Even by controlling torque to within 1% of a prescribed value, variations in the coefficient of friction can cause differences in the applied tension by as much as ±30%. Since these coefficients will vary considerably between plastic materials, it is wise to work from a laboratory test in making the choice of fastener size, style, and torque settings. The laboratory report should encompass at least this information. This may be possible only with the most modern laboratory facilities. In Chapter 6, Sec. 6.1.5, we suggested a simplified method of selecting a fastener diameter. Once diameter is determined, the basic data required for knowing the torque values to use are available. At least this is a starting point for the determination.

9.10 MANUFACTURERS (Portable Air-Powered Screwdrivers)

	Direct drive—air stall				
		Push-to-start feature		Adjustable clutch	
Supplier[a]	No clutch	(direct clutch)	Positive clutch	Without shut-off	With shut-off
The ARO Corp.	x	x	x	x	x
Black & Dekker Air Tool Div.	x	x	x	x	x
Chicago Pneumatic Tool Co.	x	—	x	x	—
Desoutter, Inc.	x	x	x	x	x
Dresser Industries Industrial Tool Div.	x	x	x	—	x
Gardner-Denver Cooper Industries	x	x	x	—	x
Ingersoll-Rand, Power Tool Div.	x	x	x	x	x
Rockwell Int'l., Power Tool Div.	x	x	x	x	x
Sioux Tools, Inc.	x	x	x	x	x
Stanley Air Tools Division	x	x	x	x	x

[a]Supplier addresses will be found in Appendix 1.

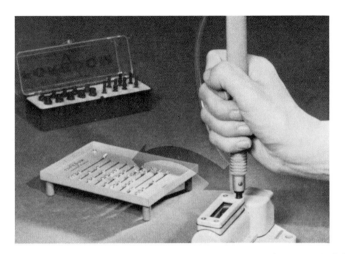

FIG. 9.6   Suction pick-up system used with small AC driver.
Operating sequence: (1) Motor and vacuum pump run continuously
and quietly.   (2) Fasteners are distributed in shaker tray, properly
oriented for pickup.   (3) Holding handpiece vertically, operator
guides it to fastener, where vacuum "draws" fastener into finder tip,
holding it ready for power driving.   (4) Operator guides handpiece
with fastener into position over workpiece.   (5) Light downward
pressure engages handpiece clutch, drives fastener to preset torque.
Clutch disengages.   (6) Operator lifts handpiece, repeats operation.
(Courtesy of The Foredom Electric Company, Bethel, CT.)

9.11   A WORD ABOUT FASTENER PICK-UP METHODS

With very small screws, the use of a magnetic or, in the case of nonferrous fasteners, a suction pick-up system on the driver should be
considered.   In one method, the screws are inserted in slotted trays
that are manually shaken to permit alignment of the heads, so that
the driver bit can easily come into contact with the recess on the
head (Fig. 9.6).   Even better, although considerably more expensive
are systems combining vacuum feed to the driver chuck and vibratory
bowls (Fig. 9.7).   In the first place, with the latter system the
screws are automatically put into position for insertion with no wasted
operator motion.   In the second place, the vibratory bowl when correctly designed and set up should be able to cull the screws being
fed to the assembly and so reduce jamming possibilities.   Although
such systems involve higher cost, 30-40% production time savings are
claimed.

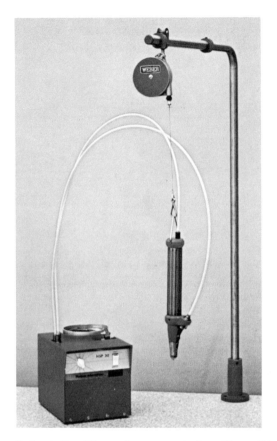

FIG. 9.7   There is no wasted motion with this vacuum-fed drive system. (Courtesy of Weber Automatic Screw Drivers & Assembly Systems, Inc. Mt. Kisco, NY.)

9.12   BENCH MODEL AUTOMATIC SCREWDRIVERS

Improved production techniques may involve the use of automatic screw driving, if justified by length of runs and cost considerations. Our concern is with the fastener specifications that will provide reliable and consistent operation in such equipment. In our view fastener requirements depend largely on the feeding systems. To illustrate we will examine two methods, both used on highly regarded automatic drivers.

In the first unit (Fig. 9.8), the fasteners are fed from a vibratory bowl, oriented to a track, and then slide down the track to an escapement. One screw at a time is escaped and loaded into the open chuck

FIG. 9.8 On this automatic screwdriver, the drive motor automatically shuts off when either the desired drive torque or depth of insertion is reached. There is plenty of space inside the chuck, and oversized heads are not a problem. (Courtesy Black & Webster, Waltham, MA.)

by means of an inserting shuttle. Contained inside the chuck jaws is a driving bit, which in the illustration happens to be of the Phillips type.

The chuck jaws then close around the screw shank, the drive bit engages the fastener's drive recess and the chuck jaws, drive bit, and screw descend toward the work piece. Rotation of the fastener then commences and the fastener is driven until appropriate seating torque or desired height is achieved.

The second unit (Fig. 9.9) also has a vibratory bowl feeder from which the screws are fed. The fasteners are then oriented and blown through a tube to a tilt-away, feed chute attached to the collet. This collet does not open; the drive chute tilts. The tilt-away, screw feed chute presents a properly oriented screw to the feed channel where it drops into the collet. The collet holds the screw in position until blade or driver bit engages the fastener. Then the unit advances to the driving position and drives the screw to the designated seating torque.

With automatic feeding and driving methods, it is easy to understand that traditional fastener manufacturing, screwhead diameter or head height tolerances, or out-of-roundness could be intolerable. The ability of a unit to accept customary screw manufacturing variables depends on the machine design, and it will pay dividends for the purchaser to

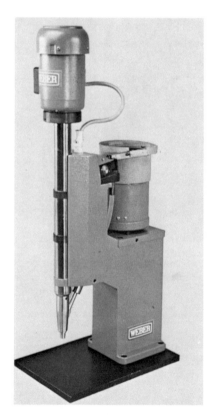

FIG. 9.9   This automatic screwdriver features a tilt-away, screw feed chute. (Courtesy of Weber Automatic Screw Drivers & Assembly Systems, Inc., Mt. Kisco, NY.)

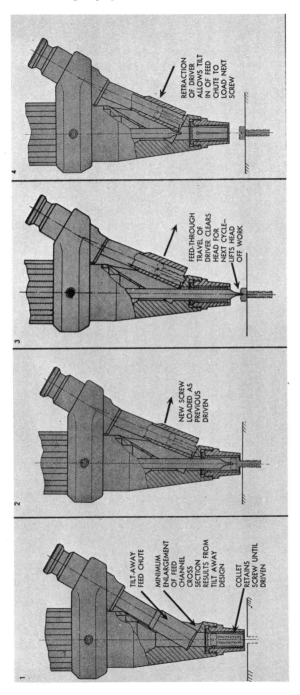

FIG. 9.10 Operation of automatic screwdriver. Screws should have length at least 0.040 in. greater than the diameter for proper operation. (From *Design News*. © 1978, Cahners Publishing Co., Inc., Boston, MA.)

carefully examine the individual features of screw handling in units under consideration.

All the automatic feeding devices we have ever seen will present some degree of difficulty when using short screws with large heads unless a highly engineered bowl system is used. Efficiency of the bowl orientation system will be lower, or there could be major problems with tumbling or jamming of screws in the tube delivery systems.

Furthermore, foreign material mixed in with the screws will not help. Of course, properly designed and set-up vibratory feeding devices will remove a great deal of junk, but they will not work as efficiently in the process.

Therefore, when automatic screwdriving equipment (Fig. 9.10) is going to be used, it is up to the user to determine what tolerances can be acceptable and so advise the fastener manufacturer when the order is placed. Installation method is also one of the questions that the fastener representative ought to inquire about.

REFERENCES

1. V. A. Williams. *Production Magazine* 90:58-63 (July 1982).
2. R. J. Finkelston and F. R. Kull. Preloading for optimum bolt efficiency. *Assembly Engineering* 17:24-28.

BIBLIOGRAPHY

Assembly Engineering Master Catalog, 1983. 21:HB-9.
Crown, P. *Modern Application News* 16:48-49 (April 1982).
Crown, P. *Modern Application News* 16:18-21 (December 1982).

10

INSERTS AND INSERTION METHODS

As noted in Chap. 2, there are special forms of nuts, called inserts (Fig. 10.1), which are frequently used to provide threaded holes in plastics, in both open and blind locations. They include nonthreaded types for lightly loaded applications as well as self-tapping, expansion, boss-cap, solid bushing, molded-in, and ultrasonically inserted versions. These inserts are generally for use with standard machine screws.

In most cases, the resultant fastening should be reserved for applications where very frequent disassembly is required, since the use of the insert represents extra cost. There will be exceptions. Exceptions include providing threads for use under considerable stress, meeting close tolerances on female threads, providing an electrical connection (1), or for difficult situations when in a bind, but no kind of insert should be thought of as a general panacea.

Inserts as illustrated in Fig. 10.1 are manufactured from brass, aluminum, steel (sometimes stainless), and some kinds from wire.

10.1 INSERT CATEGORIES

Inserts are conventionally categorized by method of insertion used. We can divide them into two major classes: (1) those that are inserted into the mold and are "molded-in" and (2) those that are installed after the molding operation is completed. These postmold inserts can also be classified in two ways:

1. Some fasteners, or fastener/inserts can be axially pushed-in to or pressed-in to the plastic part at the boss. This is done into either a taper-cored or straight-wall-cored hole, sometimes into a drilled hole.

FIG. 10.1   These representative types of internally threaded metal
inserts are but a small portion of the many variations available for use
where multiple disassemblies are required.

More or less the same conditions would prevail as with the "push-in"
fasteners covered earlier in this book, except that inserts of the types
we are now discussing have a larger exterior diameter. The insertion
method can be used in those thermoplastics and thermosets we suggested
for push-in fasteners in Chaps. 4 and 5. The press-in method can also
be accomplished under heat or by the ultrasonic method in certain thermoplastics, but not in any thermoset plastics. Expansion inserts are
also pressed-in, postmold. They come in a variety of types: flange,
clinch, wedge, cone-spread, etc. and are manufactured from brass.
With standard types, use of a special tool causes the outside of the insert to expand and anchor itself into the inside of the pilot hole.
Another kind of insert is designed to fit over the head of a boss in
the form of a cap nut. They are stamped out from thin metal and provide a single thread. Such cap nuts should eliminate or significantly
reduce thin-wall boss splitting. Because of the expense and added
operation we tend to think of them as fundamentally for use as a last
resort. When using cap nuts it is advisable that the screw exert pressure only on the nut and not exert pressure on the inner diameter of
the boss. *Helically coiled threaded wire inserts* are installed axially
with special tools; they also require tapping. Prevailing locking torque
is available with these wire inserts using standard machine screws.
Ultrasonic installation, which we will cover in more detail later in the
chapter, creates enough frictional heat at the interface of fastener and
plastic to provide a localized melt of the plastic. When it cools, the
plastic reforms and solidifies around the fastener (2).

2. Certain kinds of fastener/inserts have to be rotationally inserted. One kind, like the screws we have covered, have helical threads and are designed to form threads by either cutting them or roll-forming. We will not discuss these types in any detail because the installation conditions should be very largely the same as what has previously been written about screws.

10.2 IN-PLACE MOLDED INSERTS

We are going to jump right into the fire, so to speak, and say that the principal problem with molded-in inserts has to do with "hoop stress." The subject of hoop stress was touched on in Chap. 2, Sec. 2.4.2.

It has been suggested by highly experienced individuals that molded-in inserts require relatively thin bosses. In one article on molded-in inserts, a ratio of 1.5 to 1 between boss outside diameter (OD) and insert OD had been illustrated (Fig. 10.2) as a suggestion (1). On the other hand, the applications engineering manager of a well-known insert manufacturer recommends using a boss OD that is two to three times the insert OD. Thus, there seems to be a considerable variation in the two recommendations. Incidently, the latter suggestion is right in line with our own for use with thread-forming or thread-rolling screws (see Chap. 6, Sec. 6.1.2).

How do you reconcile the differences? Well, we are not sure that you have to. Our feeling is based on the one certainty with fastening in plastics—every application is different from the next. With molded-in inserts we become much more concerned with the effects of mold shrinkage. Also, the designer will be concerned with product cosmetics, and

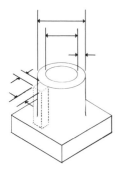

FIG. 10.2 For molded-in inserts, boss diameter should be one-and-a-half times the insert diameter. Rib at weld line can increase support. (Courtesy of E. I. DuPont de Nemours & Co., Inc., Willmington, DE.)

with thicker bosses there will be a tendency toward creating sink
marks on the exterior of the piece.

The selection of boss size should be an important concern with any
fastener and especially with in-place molded inserts in unmodified, non-
filled polymers. The reason for this concern is that the molding process
can significantly increase the possibility of delayed boss cracking. Jim
Crate, senior design engineer with the Polymer Products Department
of DuPont wrote about this subject recently*:

> The most common complaint associated with insert molding is delayed
> cracking of the surrounding plastics because of molded-in hoop
> stress. The extent of the stress can be determined by checking a
> stress/strain table for the specific material. To estimate hoop
> stress, assume that the strain in the material surrounding the in-
> sert is equivalent to the mold shrinkage. Multiply the mold
> shrinkage by the flexural modulus of the material (shrinkage times
> modulus equals stress). A quick comparison of the shrinkage
> rates for nylon and acetal homopolymer, however, puts things in
> better perspective. Nylon, which has a nominal mold shrinkage
> rate of 0.015 inch/inch, has a clear advantage over acetal homopoly-
> mer, with a nominal mold shrinkage rate of 0.020 inch/inch, in ap-
> plications requiring a molded-in insert where the resins are equally
> well suited. The higher rate of shrinkage for acetal homopolymer
> yields a stress of approximately 7600 psi, which is about 75 percent
> of the ultimate tensile strength of the material. The thickness of
> the boss material surrounding an insert must be adequate to with-
> stand this stress. As thickness is increased, so is mold shrinkage.
> If the useful part life is 100,000 hours, the 7600-psi stress will be
> reduced to approximately 2150 psi. While this normally would not
> appear to be critical, long-term data on creep (derived from data
> for plastics pipe) suggest the possibility that a constant stress of
> 2600 psi for 100,000 hours will lead to failure of the acetal homo-
> polymer part. If the part is exposed to elevated temperatures, ad-
> ditional stress, stress risers or an adverse environment, it could
> easily fracture. Because of the possibility of such long-term failure,
> designers should consider the impact grades of acetal when such
> criteria as stiffness, low coefficient of friction and spring-like prop-
> erties indicate that acetal would be the best material for the particu-
> lar application. These grades have a higher elongation, lower mold
> shrinkage, and better resistance to the stress concentration in-
> duced by the sharp edges of the metal insert (1).

---

*From Ref. 1, quoted by permission of the publisher and E. I. DuPont
de Nemours & Co., Inc.

You might feel that we are trying to discourage selection of in-place insert molding as an application method. That is not meant to be the case, although we suspect that the molders would have you prefer post-mold insertions. However, circumstances will arise where molding-in is the only way to go and the added cost may be justified.

10.2.1 Design of In-Place Molded Inserts or Studs

Because of added stress considerations in the molding process, design of any molded-in metal stud or insert should eliminate sharp corners and the bottom should be rounded (Fig. 10.3).

Shallow depths of material between the bottom of the insert and the opposite material surface will encourage sink marks and, in addition, should be assiduously avoided if there is an opposing insert. It has also been recommended that the bottom of the insert or stud not coincide with the base of the boss (2).

In any case, perhaps because of the lengthened mold cycle times and the potential for costly mold damage, the direction of interest in regard to inserts seems to be toward postmold assembly.

10.3 EXPANSION INSERTS

Expansion inserts are available in two main types: (1) cone spread, one-piece inserts (Fig. 10.4) and (2) two-piece units having a cylindrical

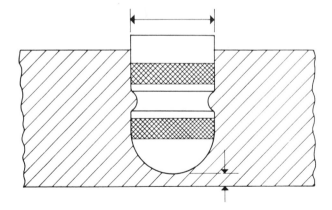

FIG. 10.3 Bottom of insert should be rounded and sharp corners avoided for in-place molded-in inserts. (Courtesy of E. I. DuPont de Nemours & Co., Inc., Willmington, DE.)

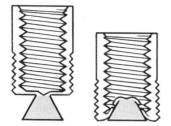

FIG. 10.4   Cone-spread inserts are
joined to a cone-shaped base. When
insert is pushed into the hole in the
boss, the cone is forced up into the
insert and expands it. (Courtesy of
Heli-Coil Products, Div. of Mite
Corporation, Danbury, CT.)

spreader that is forced down into position inside the bottom of the
internally threaded insert, thus expanding it (Fig. 10.5). Expansion
inserts are installed in straight wall holes. They are manufactured
from brass and primarily used in the electrical/electronic industries.
When cone-spread inserts are forced into a bottoming hole, the cone is
forced upward inside the insert, spreading it out and anchoring the
insert in the parent material. Since the cone is now inside of the insert bottom, the machine screw used with it must be shorter than the
length of the insert.

   Two-piece expansion inserts are available in a variety of styles:
standard, flange, wedge, and clinch styles. In all cases, introduction
of a special shape insert tool is required to spread open the cone-shaped,
slotted and knurled, or barbed bottom, anchoring the insert into the
parent material. Standard flange and clinch types are well suited to
thin wall installations. The wedge type is useful in highly ductile plastics (and wood).

10.4   CAP NUTS

Cap nuts (Fig. 10.6) are designed to allow higher torque with thin-wall
bosses. As can be seen in the illustration, they fit over the top of the

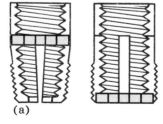

(a)

(b)

FIG. 10.5   Standard, flange, clinch, and wedge-type inserts consist
of a (a) threaded insert and (b) captivated spreader plate that can be
forced down into the slotted portion of the insert, expanding it. (Courtesy of Heli-Coil Products, Div. of Mite Corporation, Danbury, CT.)

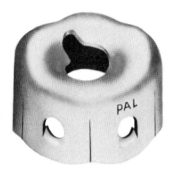
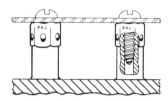

FIG. 10.6 Stamped Onsert® fasteners fit over top of bosses. For use with machine screws, standard or special thread-forming screws for plastics or type F thread cutters, such cap nuts reduce or eliminate the tendency of bosses to split. (Courtesy of Carr Division of TRW, Inc., Cambridge, MA.)

boss. They are for use with machine screws or could be used with standard type F thread cutters. The use of cap nuts adds an extra operation, but the nuts can be very helpful in avoiding problems where the boss may be too fragile.

10.5 HELICALLY COILED WIRE THREAD INSERTS

Helically coiled wire thread inserts are manufactured from 18-8 (300 series) stainless or phosphor bronze when there are salt water applications, and also if electrical conductivity or low permeability is required. Round wire is rolled to a diamond shape so that the ID and OD surface of the wire complies with the standard 60° thread of a machine screw. The inserts are provided either with free-running threads or with a locking thread on one or more coils. To produce the locking effect, the coiled wire curve is flattened into a series of polygonal "chords."

Considering all the types of available inserts, probably the use of wire inserts (Fig. 10.7) will provide the highest quality fastening sites. The reason lies in the somewhat forgiving nature of the coiled wire, which permits an even load distribution (a highly desirable feature with plastics). Furthermore, among all insert types these accept the greatest torsional and tensile strengths. Helically coiled wire thread inserts are provided in different lengths: 1, 1½, 2, 2½, and 3 times the nominal screw thread diameter. With insertions into metal, this allows the bolt tensile to be matched against the shear strength of the nut member or parent material of the assembly so that the bolt can break first.

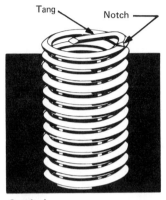

Standard

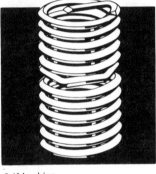

Self-Locking

Standard inserts feature smooth, free-running threads for all inch and metric sizes

Self-locking inserts have a series of cords on one or more coils

FIG. 10.7   Types of helically coiled wire thread inserts.   (Courtesy of Heli-Coil Products, Div. of Mite Corporation, Danbury, CT.)

   The originating manufacturer of this type of insert has charts comparing the tensile strength of the insert assemblies to the shear strength of several aluminum grades, magnesium and 1010 steel by insert length, but has not done so with plastics. Studies could be done in plastics, most usefully with reinforced engineering grades, but it would be an extremely lengthy job.

10.5.1   Insert Fastening–Site Design

As a general rule of thumb for these coiled wire inserts with plastics, an insert engagement length of two screw thread diameters is recommended. In regard to boss diameter versus nominal bolt size a ratio of 2:1 is recommended for helically coiled wire inserts that go into tapped holes. Critical applications may require using the ratio as against the major diameter of the tap.

The design of the fastening site is similar to that used with screws, except that the diameter of the tapped hole for use with inserts would naturally be larger than that required to accept screws only—in fact, it will be larger than the clearance hole in the member being joined to the parent material. For tapping purpose, the pilot hole will be straight sided.

In repose, before insertion, a coiled wire insert has a larger diameter than the tapped pilot hole. The insert is then prewound (to a smaller diameter) and helically inserted into the tapped hole by means of an inserting tool. The inserting tool is provided with a contoured gripping edge at the tip which engages a driving tang on the bottom of the coiled insert. When torque is applied either by hand or with an air- or electrical-powered hand tool (as mentioned in Chap. 9, Sec. 9.2.1) the wire coil can be wound into the tapped hole, following the tapped threads. On removal of the inserting tool, the wire insert will press outward against the inside of the threads in the pilot hole. Then the wire-driving tang which is notched can be broken off with another special tool.

Thus, the retention effect of helically coiled wire thread inserts extends over their entire engagement lengths. At the same time, however, a really significant advantage of this type of insert used in plastics is the lack of stress present in the assembly.

Laboratory testing techniques (Fig. 10.8) are very similar to that used with screws. Tests for tensile pullout, drive torque (insert rotation torque), and clamping torque testing are the same. What we called

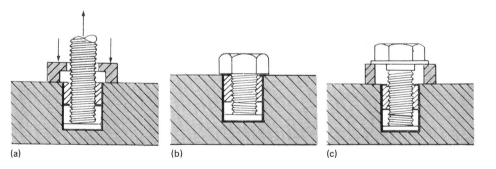

(a)   (b)   (c)

FIG. 10.8 Testing techniques. (a) Tensile strength. Axial force (in pounds) required to pull the insert out of the material at least .020 in. (b) Rotational torque. Rotational force required to turn the insert in the parent material. It is a good comparative measure of overall strength of the assembly. (c) Jack-out torque. Rotational force (inch/ounces, inch/pounds, or foot/pounds) applied to a mating screw which pulls the insert out of the material through a washer with adequate clearance for the insert outside diameter. (Courtesy of Heli-Coil Products, Div. of Mite Corporation, Danbury, CT.)

"strip torque" in testing screws is different because the insert manufacturer's laboratory will test the amount of torque applied to the mating
screw that is required to "draw out" the whole insert from the pilot hole.
This is called "jack-out torque" and results from failure of the material.

In testing inserts, if the material fails before the bolt, we will ques-
tion whether there is sufficient engagement length; with inserts it is
desirable for the bolt to fail first, a result which in plastics will depend
on the structure of the plastic and its ability to withstand shear. If
the designer is going to use coiled wire inserts, then the costs and extra operation of thread tapping must be considered.

10.5.2 Tapping in Plastics

In Chap. 2, Sec. 2.2.6, we briefly alluded to this subject in discussing
plastic fasteners for use in plastics, and we indicated that tap selection
for plastics should consider the elasticity of the plastic. The subject
of tapping is also discussed in some length in Chap. 6.

10.6 ULTRASONIC INSERTION

Ultrasonic insertion is another method of providing low stress assemblies. The field of ultrasonics came into being following World War II.
The original developments were for degreasing, cleaning, and detection
of material flaws. There were even medical applications, connected with
muscle relaxation. Later, with a constant series of improvements up
through the 1970s, ultrasonic techniques were developed for welding,
staking, and forming of thermoplastics (even sewing of synthetic material blends). Of most interest to us has been the introduction of ultrasonic techniques for inserting fasteners into plastics. Ultrasonic
techniques are used with thermoplastics only.

The many advantages of ultrasonic insertion include speed of assembly,
reduced mold fabrication and repair costs, and low insertion stress.
The force levels used in ultrasonic insertion are low, so that fracturing
and residual stresses are not induced by the insertion. This indicates
that ultrasonics can be particularly valuable in the more fragile applications. A good usage would be in tape cassettes (Fig. 10.9).

The term "ultrasonics" refers to the fact that vibrational frequencies
above 18 kHz are inaudible (or at least largely so) to the human ear.
However, in some applications such as inserting metal fasteners into
plastics, there may be a somewhat objectionable sound caused by har-
monic vibration. Use of suitable ear protection may be called for.

10.6.1 Ultrasonic Equipment

In ultrasonics, conventional electrical power from 150 to 3200 watts is
converted into mechanical vibration using a piezoelectric transducer at

FIG. 10.9   An ideal nonstructural assembly using standard press-in insertion methods for special-design, helically threaded push-in fasteners for plastics. Ultrasonic insertion methods used with this type of fastener will increase the performance characteristics of the fastener and should make possible structural assemblies.

20 kHz. Since this is at low-amplitude mechanical output, the output can be varied by means of a booster horn or mechanical impedance transformer.

The function of the "horn" is to increase or reduce (generally, reduce for fastener insertion) the amplitude of the vibration and couple it to the piece (the fastener or whatever is being joined).

The exterior face of the horn will be either a flat surface or concave, or will conform to the top of the head of the fastener and will be coupled to the top of the fastener. For other uses, the shape of the horn will be different, conforming to the particular assembly.

The fastener will be partially set into a premolded, straight- or taper-cored hole or a drilled hole in the plastic or metal-plastic assembly. When the horn comes into contact with the top of the fastener, vibratory energy is transmitted through it into the thermoplastic nut member causing frictional heat. This results in a localized melt in the immediate area around the fastener and allows penetration of the fastener into the material. Vibration is terminated instantly when the current shuts off and removal of the horn from the assembled part permits the plastic to cool and re-form tightly around the flutes, shank, or body of the fastener.

In some cases, small fasteners (perhaps best with gimlet points) can be imbedded in the plastic nut member without the need or a preformed pilot hole. Some extrusion will occur. In this case, counterbore!

Ultrasonic equipment can be mounted on a stand or pneumatically actuated bench press provided with an adjustable stroke and variable clamping pressure (Fig. 10.10). Hand-held ultrasonic units are also available (Fig. 10.11).

Either type of ultrasonic system can be utilized for installing threaded fasteners, inserts, and similar components into cored or drilled holes in thermoplastics. Hand-held systems, however, rely on operator dexterity and expertise to produce similar repeatable results because of potential pressure variation (down force) and cycle time. The actual depth of insertion often depends on the operator's visual examination

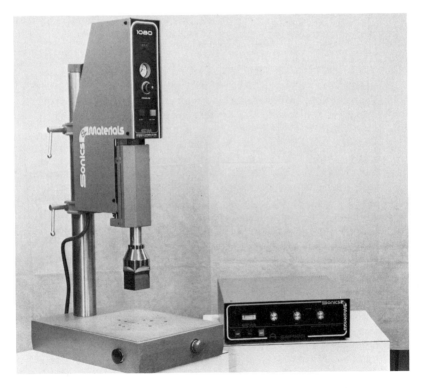

FIG. 10.10   A standard ultrasonic assembly system which can be utilized for a wide variety of ultrasonic installations in thermoplastics. (Courtesy of Sonics and Materials, Inc., Danbury, CT.)

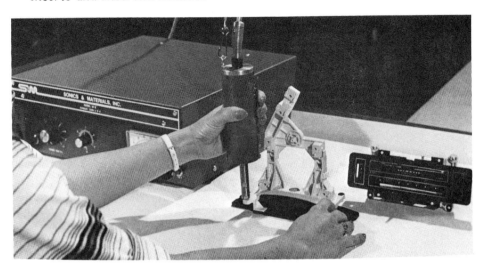

FIG 10.11 Compact hand gun for ultrasonic installations of inserts or push-in threaded fasteners. Especially useful for assembling major appliance subassemblies, furniture, ductwork, auto instrument clusters, and other large and cumbersome parts manufactured from thermoplastic materials. (Courtesy of Sonics and Materials, Inc., Danbury, CT.)

of the assembly during installation. However, bench-press type ultrasonic assembly systems include air-pressure regulators to insure consistent pressure, sophisticated electronic timer circuits to provide repeatable cycle times, and a positive mechanical stop to limit stroke length. Accessory items include pretrigger switches, lower-limit switches, micrometer adjustable depth switches, electronic depth of stroke controls, and dual-pressure systems which can insure precise depth of insertion for normal fastener or insert manufacturing tolerances.

For the push-in type screws, reference must be made to the complete ANSI standard to obtain the head height tolerances for each type of head, since this information is not covered in the manufacturer's brochures nor in the IFI "Fasteners Standards" at present.

(a)

(b)

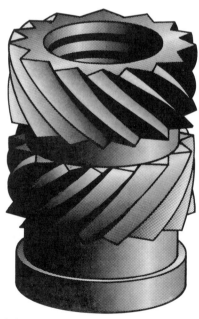

(c)

10.6.2 Special-Design Inserts for Ultrasonic Installation in Thermoplastics

There seem to be two *major* types as shown in Fig. 10.12. Type 1 is manufactured from brass. Type 1 looks remarkably similar to the trilobular "push-in" screw-type fasteners we have previously discussed, except that there are four vertical flutes and they have stepped, inclined annular ribs instead of a helical, semibuttress thread.

Type 2, or variations on it, are manufactured from either brass or aluminum. These have knurled flanges at or near the top of the insert. When provided with a ring top, the design is intended to prevent upward extrusion of plastic.

The assembly conditions and material usages of such internally threaded brass inserts should be very similar to trilobular, push-in, screw-type fasteners. Insertion should be in pilot holes with an 8° included angle taper extending to a depth as defined in the manufacturer's catalogs. Optimum use is for multiple disassembly applications.

Ultrasonic insertion is well suited to use with rivet-type fasteners with annular or helical threads, studs, and the kind of inserts previously mentioned. Selection of the type of fastener will depend on the finished product's anticipated usage, the material, and the degree of required disassembly.

10.7 SUMMARY: INSERTS FOR FREQUENT DISASSEMBLY USAGE AND PUSH-IN FASTENERS

The use of inserts will depend on the necessity for multiple disassembly, as has been stated previously. They do involve higher assembly costs.

To summarize requirements and results that can be expected with inserts and other fasteners that may be installed like inserts, we conclude with the following chart.

FIG. 10.12. Dodge ultrasonic inserts are two series of brass inserts specifically designed for ultrasonic installation in thermoplastics. (a) Type 1. (b) Type 2. (c) Variation on type 2 includes opposing diagonal knurls, an undercut and a lead-in pilot. (Courtesy of Heli-Coil Products, Div. of Mite Corporation, Danbury, CT.)

Summary: Inserts and Push-in Fasteners

Type insert	Installation requirements	Installed insert		Special remarks on material
		Holding power	Cost or speed	
Internally threaded for installation in the mold	Threaded core pins and flash removal	Excellent	Much higher molding cost	Avoid high shrinkage polymers to reduce stress
Postmold installation				
Internally threaded for ultrasonic installation	Ultrasonic equipment	Excellent	Fast	Thermoplastics only; creates very little installed stress
Internally threaded with exterior helical threads:				
Standard cutting threads on exterior	Screwdriver	Low to fair	Fast	Use for same materials recommended for std. thd. cutting and forming screws
Standard forming threads on exterior	Screwdriver	Low to fair	Fast	
Helically coiled wire thread	Taps and tapping equipment, insert mandrels and cut-off tool	Excellent	Several operations	Creates very little stress on installation
Internally threaded expansion-type	Axial press equipment and spreader tool	Excellent	Two operations	Improper installation creates stress

Stamped cap nut for press-type installation	Axial press type equipment	Fair to good	Fast	Can salvage thin-wall boss installations
Internally threaded for press-type installation	Axial press type equipment	Fair to good	Fast	Recommend for non-structural assemblies only
Special-design push-in, screw-type fastener with helical threads:				
Standard press-in installation	Axial press type equipment	Fair to good	Fast	Nonstructural assemblies; see recommendations in Appendix 3
Ultrasonic installation	Ultrasonic equipment	Excellent possible	Fast	Thermoplastics only; structural applications possible.

REFERENCES

1. J. H. Crate, Molded-in inserts: Precautions and guidelines, *Plastics Design Forum* 6:61-64 (November/December 1981).

2. A. Mussnug, Joining plastics using metal inserts, Helicoil Products, Div. IN 1061 1/80. Mite Corp., Danbury, CT.

BIBLIOGRAPHY

Miska, K. H., Joining plastics: A basic guide to the basic processes, *Materials Engineering* 88:43-46 (September 1978).

Tobias, T. J., Choosing the right postmolded insert, *Plastics Design Forum* 7:80-83 (January/February 1982).

TRADENAMES AND MANUFACTURERS

TRADENAMES

Special-Design Thread-Rolling Screws for Plastics

Plascrew®:* A trademark of AKKO Fasteners, Inc., 6855 Cornell
Road, Cincinnati, OH 45245

Plastite®: An original trademark of Continental Screw (now
Continental/Midland, 25000 Southwestern Ave., Park Forest,
IL 60466
Refers to a family of trilobular screws for plastics, manufactured
by Continental/Midland and more than 40 manufacturing licensees,†
as shown later in this appendix.

Plasform®: A trademark of Elco Industries, Inc., 1111 Samuelson
Road, Rockford, IL 61101

PLT-30®: A trademark of Industrial Fasteners Corp., 7 Harbor
Park Drive, Port Washington, NY 11050

Polyfast®: A trademark of NL Southern Screw, P.O. Box 1360
Statesville, NC 28677

NTS®: A trademark of Reed and Prince Mfg. Co., 100 Fitzgerald
Drive, Jaffrey, NH 03452

Hi-Lo®: A trademark of Illinois Tool Works, Inc., Shakeproof Div.,
St. Charles Road, Elgin, IL 60120

---

*These screws may be purchased with type 25 cutting flutes.

†One of the licensees, Camcar/Textron, manufactures a single-lead,
48° Plastite screw. Otherwise, the Plastite screws covered in this book
are manufactured as either a double-lead, 48° screw or a single lead,
45° screw.

Other Tradenames

BF-Plus®: A trademark of Camcar/Textron, 516-18th Ave., Rockford,
IL 61101
  Covers a high-quality BF-type thread-cutting screw mentioned
  in Sec. 2.3.3.
On-Sert®: A trademark of TRW, Inc., Palnut Division, Glen Road,
Mountainside, N.J. 07092
  On-Sert fasteners are stamped cap-type nut inserts that fit over
  the top of plastic bosses for use with machine screws, thread-
  forming screws or thread cutters.
Pozi-Driv®: A trademark of Phillips Screw Co., One Kondelin Rd.,
Gloucester, MA 01776
  The Posi-Driv trademark covers an improved version of the
  Phillips recess and is widely licensed in the United States and
  internationally.
Plasgrip®: A trademark of Elco Industries, Inc., 1111 Samuelson
Road, Rockford, IL 61101
Presgrip®:† A trademark of Akko Fastener, Inc., 6855 Cornell Road
Cincinnati, OH 45242
Pushtite®:‡ A trademark of Camcar/Textron
Supadriv®: A trademark of GKN Fasteners Limited, which covers a
  patented recess that is widely licensed to fastener manufacturers
  in the United States and internationally.
Torx®: A trademark of Camcar/Textron covering a patented recess
  that is widely licensed to fastener manufacturers in the United
  States and internationally.
Twin-Tite®: A trademark of Continental/Midland

**Manufacturers and Licensees of Special-Design
Screws for Plastics**

Plascrew®: Akko Fasteners, Inc., Cincinnati, OH
Jet Fastener Corp., Elk Grove Village, IL

---

*
 Plasgrip fasteners are round-body, axially inserted push-in fasteners
with a helical, buttress-type thread form.

†Presgrip fasteners are axially inserted push-in type fasteners.

‡Pushtite fasteners are trilobular, axially inserted push-in fasteners
developed by Camcar/Textron with a helical, 55°-5°, semibuttress
thread form. They are also manufactured by Continental/Midland with
a 70°-10° semibuttress thread form as Pushtite II fasteners and are
available through the trilobular fastener manufacturing licensee program.

Plastite® and Pushtite®
Australia
    Sidney Cooke Fasteners Pty. Ltd., East Brunswick, Victoria
    W. A. Deutsher Pty. Ltd., Moorabbin, Victoria
    Spurway's Industries Pty. Ltd., Alexandria, N.S.W.
Brazil
    Brazaco-Mapri Industrias Metalurgicas, S. A., Sao Paulo
Canada
    Linread Canada Limited, Guelph, Ontario
England
    Bulten-Newall, Helensburgh, Dunbartonshire
    GKN Screws and Fasteners Ltd., Smethwick, Warley, W. Midlands
    Glynwed Screws and Fastenings Ltd., Darlaston, Wednesbury,
        Staffs
    Linread Limited, Birmingham
France
    Gobin-Daude, S.A., Bonneuil Sur Marne
    Produits Trefiles de la Bridoire, La Bridoire
    Rivex, S.A., Paris
Italy
    Officine Meccaniche Pontine "O.M.P.," Milano
Japan
    Japan Fastener Research & Engineering Co. Ltd., Tokyo
    Aoyama Seisaku Sho Co. Ltd., Aichi
    Isogai Byo-Ra Kogyo Co. Ltd., Tokyo
    Kamiyama Tekko Sho Co. Ltd., Osaka
    Naito Sei Byo Sho Co. Ltd., Tokyo
    Nitto Seiko Co. Ltd., Kyoto
    Owari Seiki Co. Ltd., Nagoya
    Saga Tekko Sho Co. Ltd., Saga
    Topura Co. Ltd., Kanagawa
    Yamashina Seiko-Sho Co., Ltd., Kyoto
Republic of China
    Shi-Ho Screw Industrial Co. Ltd., Kaohsiung, Taiwan
South Africa
    National Bolts Limited, Boksburg North, Transvaal
Spain
    G. Echevarria y Cia., S.A., Renteria, Guipuzcoa
Sweden
    Bulten AB, Hallstahammar
Switzerland
    Von Moos Stahl AG, Luzerne
West Germany
    L. & C. Arnold GmbH, Forchtenberg Ernsbach
    Friedr. Boesner GmbH, Neuwied
United States
    Anchor Fasteners, Waterbury, CT
    Camcar/Textron, Rockford, IL

[United States]
   Central Screw, Keene, NH, and Frankfort, KY
   Continental/Midland, Park Forrest, IL
   Elco Industries, Rockford, IL
   The E.W. Ferry Screw Products, Inc., Cleveland, OH
   Medalist-Champion Screw, Chicago, IL
   Pheoll Mfg. Co., Chicago, IL
   Russell, Burdsall and Ward Corp., Mentor, OH
   Shakeproof Div. ITW, Elgin, IL

MANUFACTURERS/SUPPLIERS LISTED

Portable, Air-Powered Screwdrivers

The ARO Corp
One Aro Center
Bryan, OH 43506

Black & Dekker
Air Tool Division
6225 Cochran Rd.
Solon, OH 44139

Chicago Pneumatic Tool Co.
2200 Bleecker St.
Utica, NY 13503

Desoutter, Inc.
11845 Brookfield
Livonia, MI 48150

Dresser Industries
Industrial Tool Division
P.O. Box 40430
7007 Pinemont
Houston, TX 77040

(Gardner-Denver)
Cooper Air Tools
P.O. Box 1410
666 Allis Chalmers Road
Lexington, SC 29072

Ingersoll-Rand
Power Tool Division
28 Kennedy Blvd.
East Brunswick, NJ 08816

Rockwell International
Power Tool Division
400 N Lexington Ave.
Pittsburgh, PA 15208

Sioux Tools, Inc.
P.O. Box 507
2901 Floyd Blvd.
Sioux City, IA 51102

Stanley Air Tools Division
700 Beta Drive
Cleveland, OH 44143

Electric-Powered Screwdrivers

Desoutter, Inc.
(See Portable Air-Powered Screwdrivers)

The Foredom Electric Co.
Bethel, CT 06801

Jergens, Inc.
19520 Nottingham Road
Cleveland, OH 44110

Weber Automatic Screw Drivers and Assembly Systems, Inc.
P.O. Box 577
45 Kenisco Drive
Mt. Kisco, NY 10549

Bench Model Automatic Screwdrivers

Weber Automatic Screw Drivers and Assembly Systems, Inc.
(See Electric Powered Screwdrivers)

Black and Webster, Inc.
281 Winter St.
Waltham, MA 02254

Other Products

Inserts

Heli-Coil Products Division
Mite Corp.
Shelter Rock Lane
Danbury, CT 06810

P.S.M. Fasteners Limited
Longacres, Willenhall
West Midlands WV WV13 2JS
England

Ultrasonic Insertion Equipment

Branson Sonic Power Company
Eagle Rd.
Danbury, CT   06810

Sonics and Materials, Inc.
Kenosia Boulevard
Danbury, CT   06810

Laboratory Equipment

G.S.E. Corporation
23640 Research Drive
Farmington Hills, MI   48024

Gould, Inc.
Instrument Research Div.
3631 Perkins Ave.
Cleveland, OH   44114

appendix **2**

SOME QUESTIONS TO ASK ABOUT
FASTENER APPLICATIONS IN PLASTICS

1. What is the nut material and the material to be joined?
   a. What specific polymer or polymers to be used?
      (1) Reinforced or unreinforced?
      (2) Structural foam or RIM (RRIM) process?
      (3) Samples of materials or assembly available?
   b. How joined?
      (1) Plastic to plastics?
      (2) Metal to plastic?
      (3) Plastic to metal?

2. Who is the resin supplier? Grade number or designation of polymer known?

3. What kind of application?
   a. Structural?
   b. Under tension? How much required?
   c. Shear?
   d. Positional only (nonstructural)?
   e. Are there cosmetic requirements on application? Where?

4. New design or existing design?
   a. Existing—what are details of fastener and fastener site.
      (1) Engagement length?
      (2) Hole size?
      (3) Boss diameter?
   b. New design—what are the space limitations at fastening site?

5. Environment of application.
   a. Hostile?
   b. Why?

6. Frequent disassembly? How often?

7. Type of fastener required.
    a. Insert type?
    b. Rivet type?
    c. Screw type?
    d. Any fastener finish requirements?

8. What about the fastener assembly method?
    a. Molded-in?
    b. Helical insertion?
    c. Push-in method?
        (1) Axial press (only)?
        (2) Thermal insertion?
        (3) Ultrasonic insertion?

9. What driving method?
    a. Hand held?
    b. Fixtured?
    c. Automatic feed? Multiple driving?
        (1) Describe feeding method.
        (2) Any requirements relative to fastener caused by
             equipment used?

10. Quantity?

SUMMARY: FASTENER RECOMMENDATIONS

Material (molding compound)	Special-design thread-rolling fasteners	Standard thread-forming screws and inserts
Thermoplastics (crystalline)		
Olefins		
Polyethylene (PE)		x
Polypropylene (PP)		x
Polyallomer (PA)		x
Glass-filled Olefins	x	
High performance (crystalline)		
Acetal, homo- and copolymer	x	
Cellulosics		
Cellulose proprionate (CP)	Q/A	
Cellulose butyrate	Q/A	
Fluoropolymers		
TFE/PTFE and FEP	Q/A	
Nylon (polymide)	x	
Polyester (PBT/PBT)	x	
Polyphenylene sulfide	x	
Thermoplastics (amorphous)		
Vinyls		
PVC (rigid)		x
PVC (reinforced)	x	
High performance (amorphous)		
Acrylic	x	
Polycarbonate (PCO)	Q/A	
Polyetherimide	x	
Polyimide (PI)	x	
(poly)Phenylene oxide (Mod.)	x	
Styrenics:		
Polystyrene (PS)	x	
ABS	x	
SAN	x	
Sulfones:		
Polyethersulfone (PES)	Q/A	
Polysulfone	Q/A	

Standard and special-design thread-cutting screws and inserts	Axially pressed-in, push-in screws	Ultrasonically installed push-in screws and inserts	Molded-in inserts
—	x	Q/A	Q/A
—	x	Q/A	Q/A
—	x	Q/A	Q/A
	x	x	x
	x	x	Q/A
x	Q/A	x	x
x	Q/A	x	x
	—	Q/A	—
	x	x	x
x	x	x	Q/A
	—	x	x
	x	x	x
	x	x	x
	Q/A	x	x
x	Q/A	x	Q/A
	—	x	Q/A
	x	Q/A	x
x	x		
	x	x	x
	x	x	x
	x	x	x
	—	x	x
	—	x	x

Material (molding compound)	Special-design thread-rolling fasteners	Standard thread-forming screws and inserts
High-performance thermoplastic (reinforced)	x	
Structural foams	x	
Thermoset plastics		
Diallyl pthalate (DAP)	Q/A	
Formaldehydes (amino resins)	Q/A	
Phenolic	Q/A	
Polyester (thermoset)	x	
Urethane foam (if spongy)	—	—
RIM (RRIM) process	x	

x   =   satisfactory-to-excellent applications can be obtained.

Q/A =   qualified applications can be obtained.

—   =   applications not recommended.

Standard and special-design thread-cutting screws and inserts	Axially pressed-in, push-in screws	Ultrasonically installed push-in screws and inserts	Molded-in inserts
Q/A	Q/A	x	Q/A
—	Q/A	x	—
x	—	—	x
x	—	—	x
x	—	—	x
Q/A	—	—	Q/A
—	—	—	—
—	—	—	—

INDEX